Oil and the Congressional Process

Oil and the Congressional Process

The Limits of Symbolic
Politics

Bruce Ian Oppenheimer
Brandeis University

Lexington Books
D.C. Heath and Company
Lexington, Massachusetts
Toronto   London

Library of Congress Cataloging in Publication Data

Oppenheimer, Bruce Ian.
   Oil and the congressional process.

   1. Petroleum industry and trade—United States. 2. Petroleum law and
legislation—United   States.   3. Water—Pollution—Law   and   legislation—
United States. 4. Petroleum—Taxation—United States. I. Title.
HD9566.064   338.2'7'2820973   74-312
ISBN 0-669-92734-1

Published simultaneously in Canada.

Printed in the United States of America.

International Standard Book Number: 0-669-92734-1

Library of Congress Catalog Card Number: 74-312

To My Parents
and
in Memory of Andy

Contents

List of Tables

List of Charts

Acknowledgments

Since I began this research in late 1970, I have had the good fortune to have the assistance of many individuals and organizations. The assistance has ranged from providing financial support to providing access to materials, from offering scholarly criticism to offering words of encouragement, and from people patiently listening to me joyously discussing my findings to listening just as patiently to my expressions of frustration. All of this has made my research effort considerably easier to accomplish.

I am especially appreciative of John Manley, who first interested me in the study of Congress when I was a graduate student at the University of Wisconsin. He provided important direction for my research and valuable comments on various drafts. Without his ground-breaking work on the Ways and Means Committee, my study would not have been undertaken.

Important comments on earlier drafts of this manuscript were significant in shaping this final product. Austin Ranney offered prompt, patient, and encouraging direction. His substantive and editorial comments have proven invaluable. Murray Edelman and Charles Cnudde also read drafts and helped gather together many of the loose ends. In addition, Murray Edelman's work on symbolic politics provided major theoretical underpinning for this research. He remains one of the most creative contributors to the political science discipline. Barbara Hinckley and Gerald Marwell posed important questions that resulted in refinements in the research.

Additional substantive assistance came from four people who were colleagues of mine as fellows at the Brookings Institution during 1970-1971. John Ferejohn, Craig Liske, David Seidman, and John Wanat constantly provided a sounding board for my ideas. The interchange with them was something that cannot be duplicated. The Brookings Institution provided the ideal environment for encouraging this situation. Not only did Brookings present me with fine colleagues, but with financial assistance, office space, an identification with a highly respected research institution, and the freedom to operate without interference.

In the course of my research many of the people I interviewed were extremely helpful. Most were more than willing to allow substantial time for my questions. I owe a special debt of gratitude to Tom Field of Taxation with Representation. He was generous with both his time and patience, especially in untangling some of the complexities of the tax code.

My teaching colleagues in the Politics Department at Brandeis University have been an additional source of assistance. John Elliff, Roy Grow, Mark Hulliung, and Steve Rosen have regularly offered their services in various editorial capacities. Roy Macridis has been a catalyst to my efforts. Thanks is also due to Betty Griffin and Susan Wavpotich, who have typed various stages of the

manuscript from my often illegible scratching and volunteered needed grammatical correction.

I owe many thanks to my friends, who put up with me when I was preoccupied with my research while in their company. Fortunately, most of them are still friends despite my behavior.

Three people have been of more indirect influence on this work. They are the ones who had much to do with my development as a person, and it is to them—my parents and the memory of my friend, Andy—that this book is dedicated.

Finally, I should thank the oil industry. When I began the study, I had no idea that the industry's activities would become the focus of mass public attention. My real interest then, as now, was in interest group behavior. Today's concern with the oil industry will no doubt expose more people to my ideas about interest group behavior; and for that I am appreciative.

Naturally, despite the help I received from many sources, the findings presented in this book are my own, and I shoulder full responsibility for any shortcomings.

1 Introduction

"Apollo Sets Course for Home after Perfect Lunar Takeoff" read *The Washington Post* headline on July 22, 1969. In most respects, the front page stories of the *Post* did not radically differ from those of most American news sources. All other potential lead items were pushed off the front pages and became fillers for radio and television news. At other times, these events might have attracted considerably more attention. In fact, a significant political decision, occurring as the astronauts headed toward Earth, received little attention from the news media. On page two of that same edition of the *Post*, Frank Porter recounted how the Ways and Means Committee had voted to cut the oil depletion allowance from 27½ percent to 20 percent. On the surface, the story suggested that the oil industry, reputed to be one of the most powerful interest groups on Capitol Hill, had suffered a loss. This was in itself a rarity. In the past, when people had alluded to the power of private interests on Capitol Hill and to examples of this power's being exercised, the oil industry and the depletion allowance were regularly among the first mentioned. And there appeared to be considerable evidence to support this reputation. The depletion allowance had been set at 27½ percent in 1926, and despite opposition to it from three administrations—Roosevelt's, Truman's, and Kennedy's—support for decreasing or abolishing the allowance was muted in the halls of Congress. Rather than cutting the allowance for oil, the House and Senate regularly added to the list of substances covered by depletion.

What happened to bring about a different result in 1969? Why should the Ways and Means Committee, few of whose members had ever vocally proposed to cut the depletion allowance, suddenly vote 18 to 7 to do just that? Why should Congressman Hale Boggs, a Louisiana Democrat and member of the Committee who after the vote accurately described himself as "a lifelong friend of these industries,"[1] have authored the motion that carried the Committee?

Had the industry's influence really diminished? Was its reputation unfounded? Or was the loss not really a loss after all, but only designed to look like one?

In the course of this book the answers to these questions, among others, will be obtained. But these questions, while they are of significance in and of themselves, and while the Ways and Means Committee's cutting of depletion allowances was an important point in the legislative history of the Tax Reform Bill of 1969, highlight a more fundamental point of concern for students of the American political process. That concern is over what factors affect the success

1

interest groups have in the political process. Why do some win and others lose? How much is won or lost? How much is winning or losing affected by the behavior adopted by an interest group? And, why does the same group have success with its position on one issue and fail miserably with another?

To the individual with specific interests in the oil industry and the success of its maintaining the depletion allowance, or in its position on a second issue area of concern in this book—water pollution legislation—there is much to be found here. However, the main value of this book is to the student of American politics concerned with improving our systematic understanding of interest group politics in the Congressional process. The book's primary purpose is to report on the investigation of variation in interest group behavior and success across a range of policy. The findings, it is true, are specific to the oil industry and the two issues of depletion and water pollution. But, as the reader will see, they are not limited to that narrow a focus. The answers provided in this study should be generalizable to apply to a substantial range of interest groups, issues, and political situations.

Before discussing more exactly the nature and range of this research, it is best to lay some further foundation. This can best be done through an examination of the problems that exist in the literature on interest group behavior.

The Dual Problem in Interest Group Research

The interest group literature can be divided into two major segments—case studies and general theory. An examination of each shows the limitation of its contribution to our understanding of interest group behavior.

By far the best of the interest group case studies is Bauer, Pool, and Dexter's *American Business and Public Policy*.[2] Their research on tariff legislation was the subject of numerous favorable reviews and received the Woodrow Wilson Award. Yet in the prologue to the new edition, they admit that the limited issue focus was an unavoidable shortcoming of their study. In discussing this point they refer to Theodore Lowi's review[3] of the book:

Lowi's review recognized that, in important respects, the structure of an issue and of the arena in which it is discussed shapes the way in which group conflict—which clearly does occur all the time—gets translated into a policy outcome. . . .[4]

They agree with Lowi that tariff policy in the 1920s was what Lowi would call "distributive" politics, while in the era in which they deal it was largely "regulatory" politics. And it follows that "the nature of group interactions and their effects upon the outcome are different in these two types of situations."[5]

Lowi clearly pinpointed a major problem with empirical research on interest

groups in the congressional process (as well as certain other areas of policy research): for the most part the works are case studies, which significantly limits the generalizability of their findings. The works of Schattschneider, Odegard, Herring, Gross, and Milbrath, among others, are valuable but they do not provide or cumulate to *systematic* understanding of variations in interest group activity, strategy, and success.[6]

The problem that the case studies present is compounded and not corrected by interest group theory. This situation is certainly not due to a paucity of theoretical statements. A brief enumeration of the development of interest group theory and its impact on interest group research will illustrate my point. James Madison and Alexis de Tocqueville, among others, were concerned with the behavior of groups in American politics.[7] But Arthur Bentley is usually credited with being the father of modern "group theory" of politics. In the first part of *The Process of Government*, he is distressed with earlier explanations of events couched in terms of "feelings and faculties" and "ideas and ideals."[8] Because he believed these explanations were inadequate, Bentley suggested that a group focus provided a better way of viewing political events:

The whole social life in all its phases can be stated in such groups of active men, indeed must be stated in that way if a useful analysis is to be had.[9]

Bentley then proceeded to look at group interests at various levels and branches of government. His approach enabled him to cite the importance of groups in each area of political activity, but it did little to improve his ability to explain such activity. Because everything could be interpreted in terms of groups, this method of explanation lacked criteria for falsification.[10] Group theory did, however, provide an organizing framework for the analysis of political events. By present standards, one might argue that Bentley supported a new method of coding data.[11]

Unfortunately, Bentley's suggestions were not immediately accepted. Sterile descriptions of legislative politics and some reformist works continued to dominate the literature. It was not until the late 1920s that the group approach began to have some impact on political science research. At that time, several works concerned with the activity of lobbyists in Congress were published. These included Peter Odegard's book on the Anti-Saloon League,[12] Pendleton Herring's study of special interest lobbies,[13] and E.E. Schattschneider's work on the Smoot-Hawley tariff.[14] These authors interpreted the group approach more narrowly than Bentley had intended. One wonders, when reading these studies, whether the authors were even familiar with Bentley's work. None of them discusses, indexes, or footnotes Bentley or his work.

A more likely explanation for the return to a study of politics in terms of groups was offered by Schattschneider:

Political conditions in the first third of the present century were extremely hospitable to the idea. The role of business in the strongly sectional Republican

system from 1896 to 1932 made the dictatorship of business seem to be a part of the eternal order of things. Moreover, the regime as a whole seemed to be so stable that questions about the survival of the American community did not arise.[15]

Schattschneider continued that this caused problems for the group approach at a later point in time:

The difficulties are theoretical, growing in part out of sheer overstatements of the idea and in part out of some confusion about the nature of modern government.[16]

In any case, there exists reasonable doubt over the impact of Bentley's theories on these early applications of the group approach.

Thus, while there existed both a theoretical statement and some case studies, they were on parallel lines that did not intersect. Nevertheless, Bentley's work has proved a valuable contribution to political science. Not only did he question the use of the individual as the unit of analysis, but he also was the first to consider seriously the implications of overlapping interest and other sources of cross-pressures. But it was not until the 1950s that Bentley's contribution was fully recognized. In 1953, David Easton observed that, "It is only today, indeed, that a more general and belated revival of *The Process of Government* has become apparent."[17]

Easton was referring to the works of David Truman. It was Truman who performed the major resurrection and adaptation of Bentley.[18] Truman should also be credited with a clarification of Bentley's theory. In addition, he focused attention on the valuable component of potential groups, those clusterings of individuals with shared attitudes who do not interact.[19] But the theory he posits appears to have little relevance when he applies it to interest group activity in various political arenas. For example, in his chapter on "Techniques of Interest Groups in the Legislative Process," Truman notes that, " . . . access is the fundamental objective of group activities."[20] He then provides a laundry list of examples of where and how groups obtain access. But he never discusses what causes variation in access from one policy to another, why certain groups have better access than others, or why strategy X is preferable to strategy Y. Truman provides us with the understanding of how interest groups deal with constant conditions in the legislative process, but not how variations in policy, process, or the groups involved affects the outcome. This takes us little beyond the case studies, which also may identify constants but are unable to deal with variability outside their single issue focus.

Giovanni Sartori recently commented on this dual problem in the study of interest groups:

Hence the "indefinite group" of the theory, and the "concrete groups" of research fall wide apart. The unfortunate consequences are not only that the research lacks theoretical backing (for want of medium level categories, and

especially of a taxonomic framework), but that the vagueness of the theory has no fit for the specificity of the findings.[21]

We have, then, case studies that are not generalizable and theory that has not and perhaps cannot be empirically tested. And like the man who has a large roll of dollar bills, we may deceive ourselves into thinking that the quantity of literature makes us rich, when in fact we are quite poor.

Comparative Policy Analysis: Bridging the Gap

The problem thus remains to make some connections between interest group theory and studies of interest group activity. An essential step in this direction is to get beyond the simple interest group case study and develop some generalizations based on empirical evidence.

With this in mind, I decided to undertake a comparative policy analysis. Stimulated by Lowi's discussion of *American Business and Public Policy*, I thought it important to get an understanding of interest group behavior across a policy range. Only in this manner can we build systematic knowledge of interest group behavior, if behavior is indeed systematic. To work in this direction, I decided to examine the activity and success of a single interest group on a range of issues. Concentration on one interest group is possible control for some sources of variation in interest group behavior—the resources of the interest, the way it is organized, and its size. These remain fairly constant.

The advantage comes from studying the interest group across a policy range. In this particular case we are examining only two issues, water pollution and the depletion allowance. But because, as I shall shortly explain, they are representative of broader policy types, we can generalize from the results with some confidence. This is not just an ordinary comparative case study. It is designed to provide some systematic understanding of variations in an interest group's activity and effectiveness.

The problem thus became one of selecting a policy typology or theory (as Sartori calls them, "medium level categories") that suggests variations in the nature of group activity and success and for which representative issues, in which the same interest group was involved, could be identified.

Four policy categorizations were considered for this purpose:[22] Froman's areal-segmental;[23] Lowi's distributive, redistributive, and regulatory;[24] Salisbury's modification of Lowi;[25] and Edelman's symbolic-material.[26] These four were chosen because they are all addressed to the idea of viewing policy as an independent variable affecting the policy process and were developed in the context of U.S. domestic politics.

A major failing of all the categorizations is that the categories lack mutual exclusivity. This problem of overlap is symptomatic of the retarded position of

the social sciences. To avoid this problem, I decided to look at the categories as the poles of continua rather than as nominal, discrete labels. In this way, a given issue can be seen as existing at some point on a continuum and can perhaps be ordered on that continuum in relation to other issues.

The ease with which the transition of the categories (from nominal to ordinal scales) could be made became a major criterion for selection of the policy typology. Other criteria included relevance of the categories to interest group politics, accuracy of description, and ease of application. On these bases, I selected the Edelman typology.

Before discussing the positive reasons for the selection of Edelman's typology, some mention should be made concerning reasons for rejecting the other three. Froman's areal-segmental typologies, despite his impressive presentation of it, contains difficulties.[27] Prime among these is the preclusion of policy continua development from his typology. Froman's definition of areal and segmental suggests that the policy types are discrete. Areal policy affects all residents of the community, while segmental affects only certain groups. This provides clear definition of policy types. Yet when one tries to classify policies by this scheme, an argument can be made that almost every policy is in some way areal.[28] In addition, Froman's typology was developed for use in the study of urban politics. It is questionable whether the policy distinction he makes, even if it has any meaning for the study of urban politics, has any applicability to federal policy decisions.

The Lowi and Salisbury works pose different kinds of disadvantages. While they are more applicable than Froman's to the study of federal policy making, the lack of discreteness combined with the increase in the number of policy types make them unmanageable. While the former quality may provide a desirable reason for the construction of continua, the latter makes them infinitely more complex. To exemplify, let us deal with Lowi's categories for a moment. Because there are three types rather than just two, it is necessary to order policies on three dimensions instead of just one. The diagram below illustrates the problem.

One must locate a policy area somewhere in three dimensional space to adequately represent the existing mix. Obviously ordering policies becomes an increasingly complex task, especially if the ordering is to have some meaning in terms of interest group activity. At some point the Lowi approach may be more desirable, but given the current state of the discipline, it just muddies the stream a little more. Results may be uninterpretable.

Salisbury may claim that his addition of the self-regulatory category actually reduces Lowi's typology back to two dimensions. But all it really provides is another modal type, and ordering of policies would have to take place on six dimensions.[29]

Obviously, one of the main reasons for selecting Edelman's symbolic-material framework is its relative simplicity. Handling the ordering of policy areas on a single dimension running from material to symbolic is a far easier task than working in three dimensional space.

Further, he does not (as Froman does) assume that the categories are discrete. When Edelman examines the operation of symbolic politics in regulatory policy, he clearly notes that no policy area is totally symbolic. In fact, symbolic processes require "a mix" of symbolic effect and rational reflection of interests in resources, though one or the other may be dominant in any particular area.[30] In effect, he is talking about a policy continuum. Politics is never purely symbolic or purely material. The focus here, as in Edelman's work, will be on the examination of interest group activity given different "mixes." (Although at a later point in the book these categories will be treated as if they are discrete—when, for example, a particular issue is mentioned as being in either a symbolic or material policy stage—this will be done for sake of simplicity. A policy mix is still implied despite use of the discrete terms).

Other benefits are obtained by the selection of Edelman's framework. His work is, in some ways, an extension of group theory. His discussion of symbolic politics is couched in terms of the political success of various strategies. Furthermore, he formulates the following hypotheses about political quiescence in terms of levels of interest group organization:

1. The interests of organized groups in tangible resources or in substantive power are less easily satiable than are interests in symbolic reassurance.
2. Conditions associated with the occurrence of an interest in symbolic reassurance are:
   a. the existence of economic conditions threatening the security of a large group;
   b. the absence of organization for the purpose of furthering the common interest of that group.
3. The pattern of political activity represented by lack of organization, interests in symbolic reassurance, and quiescence is a key element in the ability of organized groups to use political agencies in order to make good their claims on tangible resources and power, thus continuing the threat to the unorganized.[31]

I would add the following derivative hypotheses:

1. As an issue moves from the symbolic to material policy level, symbolic reassurances will be less successful in satisfying the unorganized.
2. Use of symbolic reassurance may lead to quiescence of the unorganized. But under certain conditions it may result in arousal.

3. The continued use of symbolic rewards may eventually have material consequences.

These tentative hypotheses are the ones I will investigate and refine in the course of this study.

The Selection of Issues

Once the Edelman typology was chosen, the problem became one of finding issues representative of extremes on a material-symbolic continuum. The reason for choosing issues that appeared to be extremes on this policy was quite simple. If there exists a relationship between policy and interest group behavior, one would expect—assuming a linear relationship—that the differences in interest group behavior would be most noticeable as one moved toward the policy extremes. After examining a series of domestic issues, I selected the water pollution and depletion allowance issues. The most obvious reason for this choice was that they fit with the descriptions Edelman provides for symbolic and material policy types.

Water pollution legislation has been largely symbolic in nature. While there have been numerous new enforcement laws in the post World War II era, the levels of enforcement, until recently, have been low. Mass public interest groups have become involved. The claim is often made that the polluters are the same people who take charge of enforcement. Consequences of water pollution are not easily measured in monetary terms. All of these readily observable factors, plus other less obvious ones, seem objectively to meet the pattern of symbolism Edelman describes.

While there is a good deal of symbolism in tax legislation, and specifically with the oil depletion allowance, it is very exact as to the material consequences of its enforcement. The symbolism largely comes from the overall notion that people are actually taxed according to their ability to pay.[32] However, the tax code itself is very explicit regarding rules, penalties, and conditions of payment or, in the case of depletion, deductions from payment. Further, in spite of evasion and avoidance by many, some significant level of compliance with the code is met by most individuals, organizations, and corporations. Spot checks are made in an attempt to uncover irregularities. Non-compliance is often met with a significant level of enforcement.

Thus, upon examination of basic, objective descriptions of the two issue areas, it is reasonable to claim that these were acceptable choices for representing extremes on the symbolic-material policy continuum. Again, it is not claimed that they are pure types. Certainly there are material consequences involved in pollution legislation and symbolic components in the tax field. In fact, as will be discussed in a later chapter, within each issue area inconsistencies in the behavior

of political actors can often be seen as a change in the material-symbolic politics mix of the given issue.

Pragmatic considerations also influenced the choice of issues. In the case of pollution, the fact that it is of current interest increases the willingness of people to discuss it. Naturally, the existence of a body of literature on each of the issue areas was a stimulus for their selection. In the tax area, John Manley's book *The Politics of Finance* and Stanley Surrey's seminal article on tax reform provided a foundation on which to build.[33] The literature on water pollution is fairly extensive; while most of it does not deal adequately with the politics of legislative policy making, studies by M. Kent Jennings and James Sundquist proved of exceptional value.[34] Second, given the desire to provide preliminary texts to certain hypotheses, simplicity of design proved desirable. To deal with the problems on a broader scale would require a sacrifice of thoroughness. Third, by limiting the extent of the study, we can view the impact of policy within a controlled environment. For example, both issues were a part of some Congressional activity throughout the period since 1960, and activity with each reached a climax during the 91st Congress. The time environments are therefore similar.

My decision to examine the oil industry's activity in each of these issue areas coincided with the selection of the issues themselves. The industry has been active in both areas for a considerable period of time. Robert Engler's fascinating, if unsystematic study of the oil industry in politics provided a research base.[35] Moreover, the industry has clearly identifiable lobbying organizations in continuous operation in the Congressional process. But my main reason for selecting the oil industry was that it operated in two issue areas that fit the broader policy criteria.

There are certain sacrifices in this approach, since my generalizations will be based on only two issue areas and the operation of a single interest. As with a case study, there is room for error due to the peculiarities of the individual cases. However, this limitation is not as great as one might assume. First, I have studied the issues over a considerable time period. Thus, we are not limited to only two data points, one based on the water pollution issue and the other on the depletion issue.[36] Rather, we have a series of data points for each from which to judge sources of variation. Moreover, in Chapters 2 through 4 we will examine other variables that, according to the literature, affect the activities and success of a group in the policy process. Through a careful analysis of these variables, I expect to limit the error in generalizing from these particular issues.

Additional Areas of Investigation

The goal from this intensive examination of the oil industry's activity and success in each of these issue areas is to systematize the study of interest groups within a policy frame. While the research design maintains certain controls, there

are at least three variables or variable groups which the interest group literature identifies as having impact on a group's activity and success that fall outside the controls provided by the research design. They are the constituency variable—rules, procedures, and processes—and the nature of the opposition. We will examine each of these separately to evaluate impact on the activity and success of the oil industry in each of the issue areas and to provide the framework for understanding the impact of policy.

However, there is an additional reason for the study of these variables. While the literature identifies each of them, it does little to explain the exact nature and impact of their operation. Our examination will, hopefully, provide some of the missing linkages. In the discussion of these variables, I shall specify the particular questions to be explored with regard to each.

The Constituency Variable

In Chapter 2 we shall investigate the relationship between the number and strength of constituency ties the oil industry has at the various decision points in the legislative process and the success and behavior of the industry in each of the two issue areas.

Probably no factor has been cited more often as an explanation for the success or failure of interest groups in legislative politics and the voting behavior of legislators than constituency.

Several studies note the importance of constituency ties for understanding why interest groups concentrate their activity and are more successful in their dealings with the Senate than with the House of Representatives. Richard Fenno claims that the House tends to be more economy-minded than the Senate on appropriations bills. He ascribes this to the broader scope of group interests that a Senator, as compared to a Representative, must represent.

Many but not all members of the House represent fairly homogeneous constituencies. In some of them—mostly rural ones—the ideology of minimal government, small budgets, and frugality remains strong. And, in any event, the more homogeneous a constituency the fewer the numbers of interests which must be represented. The demands for increased funds, therefore, will be concentrated on relatively few programs, and economy expectations can safely be voiced on all the others.[37]

Lewis Froman finds the same factor as one of two major causes for behavior differences in the House and Senate.[38] (However, it is worth noting that Froman then spends the remainder of one book discussing the effects of rules and procedures on decision making. In doing so he actually provides a third factor, to some degree independent of the other two, that explains House-Senate differences.)

An even more systematic discussion may be found in John Manley's book, *The Politics of Finance.* He finds it useful to discuss decision making in terms of interest aggregation at various points in the legislative process of tax, trade, and social security bills.[39] He later concludes:

The reason the Senate does better in cases of conflict with the Ways and Means Committee is because politically Senate decisions are more in line with the demands of interest groups, lobbyists, and constituents than House decisions.[40]

While these three stand out because of their concern with House-Senate differences, they are not alone in ascribing behavior patterns to constituency related explanations. Warren Miller and Donald Stokes seminal article "Constituency Influence in Congress" demonstrates that both constituency attitude and representative perception of it play a significant role in his voting behavior, albeit of varying magnitude across policy areas.[41] Earlier the work of Duncan MacRae had found strong relationships between constituency characteristics and voting behavior of Congressmen across several issue areas.[42]

More recently two major efforts have reemphasized the importance of constituency although on slightly different lines. David Mayhew looks at strength of party voting for Congressmen from interested and non-interested constituencies in four policy areas. Although his study lacks the rigorous niceties of Miller and Stokes and MacRae, his findings reinforce theirs and are based on a broader time span.[43]

John Jackson's work is the first major attempt to apply notions of constituency to Senators' voting behavior. While it poses certain problems, which will be considered later, Jackson's study again demonstrates the significant role of constituency in legislative behavior.[44]

Other studies have investigated the relationship between constituency stakes in a program and the vote of representatives. These have ranged from such programs as defense spending to Model Cities to agricultural support programs.[45]

The obvious question one may now ask is, "If all these pieces say that constituency is important, why consider it again?" The answer rests on three basic points. First, there are some who dispute the importance of constituency interest success. Lewis Dexter, for one, notes the important functions of selective attention and perception on the part of Congressmen.[46] Charles Jones indicates that most Congressmen do not have to run on their record. As long as they are incumbents, they are very difficult to unseat.[47] Further, Miller and Stokes in their article "Party Government and the Saliency of Congress" find only 6 to 7 percent of their sample used "issue" reasons to explain their vote in the 1958 Congressional race.[48] Although none of these directly indicate that Congressmen do not try to serve interests within their constituency, they certainly question whether they have overwhelming need to do so.

Second, few of the studies consider constituency impact on the success of

interests except as demonstrated in floor votes. Of more importance may be the impact of constituency on people in charge of agenda setting (such as committee chairmen), on votes in committees, on committee recruitment, and on execution of legislation. Chapter 2 attempts to establish the potential strength an interest group has from constituency ties, how an interest group uses what ties it does have, and whether the ties are operative. In addition, in the chapter on the constituency variable, I shall refine the Fenno-Manley-Froman notion about House-Senate differences in aggregation of interests.

Third, if constituency is important for explaining differences in interest group activity and success at various decision making points, constituency must be understood and controlled for in examining the two issue areas under study here. That is the only way an adequate understanding of variations across policy areas can be evaluated.

Rules, Procedures, and Processes

Woodrow Wilson recognized that rules and procedures under which Congress operates are not without impact. In *Congressional Government* he laments the restrictiveness of House rules and procedures on the individual members.

No man, when chosen to the membership of a body possessing great powers and exalted prerogatives, likes to find his activity repressed, and himself suppressed, by imperative rules and precedents which seem to have been framed for the deliberate purpose of making usefulness unattainable by individual members. Yet such the new member finds the rules and precedents of the House to be . . . [49]

As we know, Wilson's lament formed the base of an attack on the decentralized, non-responsive decision making system in Congress, and on the way this system favored the activities of organized lobbys.[50]

Many modern day political scientists, some no doubt stimulated by Wilson's observations, have investigated this decentralized process. Some have chosen to examine the operation of individual committees and have spent parts of their studies on the degree of control the committee has in the policy area, how much influence the committee and subcommittee chairmen have, and who gets recruited to the committee.[51] Others examined policy areas with some emphasis on problems that the structure of Congress creates for policy adoption.[52] And a few efforts have been pointed directly at rules, procedures, and norms.[53]

But attention to the charge Wilson raises has been inadequate. To what extent do rules, procedures, and their application affect whose position is represented?

In Chapter 3 I shall argue that rules and procedures are not neutral. They have an effect on who wins and who loses, on where interest groups spend resources, and even on what is considered. Further, they differ from one issue to

another. The Congressional process is not just a series of constants. To the degree that operations differ in the water pollution area from the depletion area, we expect the oil industry to act differently and to achieve different success levels. Also, rules, procedures, and processes may change over time within one issue area. Some of the specifics that we will consider in Chapter 3 are the committee recruitment process, the closed rule, and variations in committee operation. As with the constituency variable, I expect that understanding the impact of these factors puts us in a better position to evaluate the influence of policy type on the industry's activity and success.

In addition to contributing to our understanding of the impact of policy on interest group behavior, the chapter on rules, procedures, and process has a more general goal. In it we will explore the ways in which the Congressional process works to favor the activities of organized private interest groups who resist changes in policy. Thus, while our concern is with variations in the ways the process operates on water pollution legislation and depletion allowance legislation, I wish to call attention to the way certain constants in the Congressional process inflict a "non-decision" bias to policy making.

Interest Group Competition

Patterns of interest group activity may differ with the degree of group competition. That is, one may discover that, in a situation where all interest groups represent just one side of an issue, circumstances may differ markedly from those in a situation where some groups are active on both sides.

As will be discussed in Chapter 4, variation in levels of opposition affects such things as the decision making point at which oil industry's efforts are most effective; the levels at which the oil industry spends resources; the types of strategies it employs; and the way in which it deals with potential groups.

Unfortunately the importance of this point has been infrequently mentioned in the work on interest groups. Most of the valuable comments have been made with reference to particular policy areas and then only indirectly. However, certain works are suggestive. James L. Sundquist, in his book *Politics and Policy*, notes the importance of the development of public interest support in each of the policy areas he considers as having a major impact on the effectiveness of organized private interests.[54] Stanley Surrey's analysis of interest group success in tax legislation finds a lack of opposition a major factor in policy outcomes.[55] But beyond these, few deal with the subject in a systematic way.

Chapter 4, aside from its contribution to the long-range direction of the study, has independent importance. In it we will not just be concerned with what the oil industry does and why that is or is not effective. Instead, much of the analysis will deal with the problems that unorganized public interests have in the Congressional process. This will include some important refinements for Mancur Olson's theory of collective behavior.

The Policy Types

The reasons for the selection of the two issue areas has already been amply elaborated. When analysis of the preceding three variables or variable groups is completed, the question will still be what differences remain unexplained about interest activity in the two issue areas and can any of the remainder be accounted for in the symbolic politics discussion Edelman developed. More broadly, the question is the one raised by Lowi over whether the structure of the issue shapes group activity and, in turn, affects the policy outcome. Does the study of policy types add anything to our understanding of interest group activity and success? And can we build some middle range generalizations about variations in interest group activity that result from policy differences?

It should be clear by now that the terms "policy" and "policy type" refer here specifically to policy content. One may argue that it is difficult, if not impossible, for policy content to affect policy processes and group behavior since the actual content is not known until after the processes are concluded. But this view ignores the fact that debate over a particular issue is done within certain policy benchmarks, limits on likely policy content, and that policy content is assumed by political actors from previous dealings with an issue. Thus, the lines of debate set the range of policy content.

A major part of the chapter on policy types will also involve evaluating the nature and impact of the policy changes in the two issue areas. Rather than assume that the passage of new water pollution legislation or the lowering of the depletion allowance indicate losses for the oil industry, we will consider the impact of legislative changes to evaluate the industry's success.

Thus, each chapter has a contribution to make to the overall direction of the research—understanding the impact of policy variation on interest group behavior. Chapters 2, 3, and 4 are designed to evaluate other variables related to interest group behavior so that sources of spurious relationships between policy type and group behavior can be minimized. In addition, each of these chapters have some independence. In the course of reaching this broader goal, it was possible to investigate several auxiliary questions concerning the nature of interest group activity in the Congressional process.

Before beginning the body of this book, the reader might find it valuable to read the comments on the methods and limitations of this study set forth in the Appendix. It includes a mention of major source materials, time period of the study, and interview techniques.

We Get By With a Little Help From Our Friends

Many of the sources of support the oil industry has had on Capitol Hill have been attributed to the ability of oil industry representatives to initiate and cultivate friendships in the House, the Senate, and the executive agencies. Thus, during the course of my study, I ran into people who expounded this point of view: "Oil doesn't have any enemies on Capitol Hill." And while this exaggeration may hold a kernel of truth, it misses the major point. That is, given the way the industry is distributed, it is an important constituent for many political decision makers. Of course, the industry may develop ties to politicians whose geographic constituency is not responsible for producing even a drop of oil. And stories are well known in the corridors of Congress about Senator Russell Long and before him Senator Robert Kerr finding campaign contributors for colleagues faced with tough re-election campaigns. But the fact remains that much of the industry's success comes from simple constituency relationships. It is the importance of these simple constituency ties to oil industry success that we will examine in this chapter.

To do this, three interrelated operations or goals will be pursued. The first is to evaluate the strength of constituency relationships possessed by the oil industry at various decision making points in the legislative process. By constituency relationship, I mean the ties existing between decision makers and the industry because oil is produced in the geographic area a decision maker represents, or ties developed from direct association of the decision maker with the industry. Later we will deal with the operationalization of this constituency relationship. To understand the industry's strength, the establishment of the level of constituency ties is an important first step.

Second, an effort will be made to see if the level or strength of these ties is related to the industry's success in the two issue areas. Where data are available, the relationship of constituency strength to success, both across time and across decision making points will be examined. We can then evaluate the independent effect on variations in constituency strength on industry success.

The third goal is to look at the impact that variation in constituency strength has on the success of the industry in our two issue areas. If we are eventually to grasp whether policy area has an impact on interest aggregation, this source of variability must be controlled.

One may view this chapter as a huge analysis-of-variance problem. The dependent variable is a measure of interest group success. Several independent effects or treatments may be affecting this variable. Time may be one;

constituency strength another; and issue area a third. In later chapters new independent variables will be added.[1]

Throughout we are limited to consideration of independent effects of each variable. It is impossible to furnish estimates of coefficients of variation or anything of that kind. The data are too soft for such rigorous purposes. Sometimes only impressions are available. Moreover, many data are missing. However, if enough pieces of the puzzle are filled in, we can obtain a comprehensible picture.

The organization of this chapter will proceed with a separate examination of oil industry constituency ties in each issue area. This includes concentration at four levels of legislative decision making: House and Senate floor; committees, subcommittees, and staff; executive agencies; and the Presidency. Naturally there may be multiple decisions made at each of these levels. These two sections will be followed by one that considers the differences between the two issue areas.

Constituency Support and the Depletion Allowance

The history of the modern-day fight to reduce the oil depletion allowance begins in 1950. The allowance permits oil producers to deduct up to 27½ percent (since 1969, 22 percent) of their gross income before subtracting other deductions in computing their taxable income. There have been several rationales offered over the years for this tax treatment. Prime among them is that once oil (or other minerals) is removed from the earth, there is that much less of it to be removed, i.e. it is being depleted. As the supply is used up, oil becomes more scarce and more difficult to discover and develop. The depletion allowance is supposed to be an incentive to encourage new exploration.[2] Combined with other features of the tax law such as the expensing of intangible drilling costs, the depletion allowance permits the oil industry to pay a relatively low level of federal taxes.[3]

The Revenue Revision of 1950 was the first of many major attempts to lower the depletion allowance that occurred from 1950 until 1969, when the allowance was cut to 22 percent. It is true that Franklin Roosevelt had attacked depletion in his veto message of the Revenue Act of 1944, in which he objected to extension of depletion allowances to certain other minerals:

Percentage depletion allowances, questionable in any case, are now extended to such minerals as vermiculite, potash, feldspar, mica, talc, lepidolite, bauxite, and spodumene.[4]

But the shot was not very direct and was neatly fielded the next day in Senate Majority Leader Alben Barkley's resignation speech in objection to the veto.

Barkley noted that the President's views "coincide with the traditional views of the Treasury," which "has always been opposed to any sort of depletion allowance for the development or marketing of minerals."[5]

In 1950, President Truman took the first direct swipe at the depletion allowance. In his tax message to Congress, Truman stated:

I know of no loophole in the tax law so inequitable as the excessive exemptions now enjoyed by oil and mining interests.[6]

The Truman attack on the depletion allowance was the first of four major attempts to alter it between 1950 and 1969. The two in between were in 1951 and 1963-1964. The former was again a direct attack by Truman, while the latter was an indirect administrative approach by Kennedy in the 1963 tax cut bill. Throughout this time, numerous minor attempts to lower the depletion allowance were undertaken. These were largely in the form of floor amendments in the Senate. Some, however, took the form of applications and redefinitions of regulations by the Treasury Department. Finally, in 1969, as part of the Tax Reform Act, the Congress agreed to cut depletion for oil and gas to 22 percent. The House, as previously noted, voted for a cut to 20 percent and the Senate Finance Committee voted to raise it back to 23 percent, after failing on a tie vote to raise it back to 27½ percent. The Senate floor then rejected several amendments to the 23 percent figure (both increasing and decreasing it), and the conference committee agreed to the 20 percent figure.

The forthcoming analysis will cover the 1950-1969 period, but will concentrate on events surrounding the major attempts at changing depletion.

Depletion: The House and the Senate

The analysis of the constituency strength of the oil industry begins with an examination of the House and Senate. While these bodies are neither the first nor the last points in the decision making process, it makes sense to begin here because of the relative high visibility of their decision making as compared to the other decision points. Here we have floor debates, public statement, and at times floor votes from which to analyze the constituency impact. (Basing analysis on only highly visible data, may, of course, lead to mistaken analysis. As will be discussed at a later point, visibility of decision making can have a major impact on the success of the industry.)

A second reason for starting with the full House and full Senate is that it will provide us with a base point for estimating constituency ties of representatives and the petroleum industry. From these, comparisons of strength level can be made to other decision points.

Identifying Oil Constituencies

The first estimation problem lies in figuring how many House and Senate members could be considered as having oil interests within their constituencies. Ordinarily, one would believe that a greater percentage of Senators would have significant petroleum interests in their districts than House members. As Fenno has argued: "Senators . . . have larger and less homogeneous constituencies than do House members. Most Senators represent a near-microcosm of the nation as a whole."[7] To a large extent, the correctness of this position depends on how an interest is concentrated and whether we should assume that the same quantity of the interest in a House district as in a Senate district implies equal constituency importance. Let us explore this problem.

If we take the petroleum industries' figures on where oil is produced as indicative of an oil interest tied to a Senator from producing states, we find that thirty-three states have oil production.[8] Thus, sixty-six Senators have oil interests in their constituencies. However, when one examines the data more carefully, one finds that less than one-half of these states have significant oil production. In 1968, for example, fifteen states produced 97.5 percent of the total United States crude petroleum. Table 2-1 lists the states and their percentage of petroleum production. At some point during the 1950-1969 period, each of these states was responsible for at least 1 percent of the United

Table 2-1
U.S. Crude Petroleum Production, 1968

	Percent
Texas	34.1
Louisiana	24.6
California	11.2
Oklahoma	6.7
Wyoming	4.3
New Mexico	3.8
Kansas	2.9
Alaska	2.0
Mississippi	1.7
Illinois	1.7
Montana	1.5
Colorado	1.0
North Dakota	.7
Utah	.7
Arkansas	.6

Source: United States Department of the Interior, *Mineral Yearbook 1969* (Washington, D.C.: Government Printing Office, 1969) Vols. I-II, p. 836.

States crude petroleum production.[9] The remaining states contributed only marginally to petroleum production. It would be inaccurate therefore to consider more than thirty Senators as having significant oil producing interests in their constituencies.[10]

Between most of the 1950-1969 period, the oil industry had an estimated potential constituency support in the Senate of thirty votes. (Prior to Alaskan statehood the potential constituency support was only twenty-eight votes.) Naturally, this remained constant, with the above exception, since each state has only two Senators.

In estimating constituency strength in the House we are faced with a different problem. For example, a ten-million dollar industry might be insignificant to a state or to a Senator from that state, but it might be extremely significant to a House member and his district. An industry that is large in a given Congressional district might make a substantial contribution to the well-being of a large number of people. Suppose that industry employs 2,000 people in a given district. Those 2,000 people are far less important electorally to the Senator from that state than they are for the Congressman from that district. (This assumes that there are multiple districts in the state.)

Establishing an oil producing interest in a Congressman's constituency requires a criterion separate from the one used for a Senator. If a definition were used that would allow us only to include Congressmen (from the fifteen leading petroleum producing states) who had petroleum production in their districts, only eighty-four House members could in 1969 be classified as representing constituencies with oil interests. Instead, evidence of oil production in the counties composing the district was used to classify it as one with or without significant oil producing interest. Thus, in a state outside the fifteen major producers it would be possible to locate a Congressional district with a petroleum interest.

For operationalizing this definition of a constituency petroleum interest *The Mineral Yearbook 1966* was used. Districts were considered to have a petroleum producing interest if (1) the value of petroleum produced in 1966 in that district exceeded five million dollars; (2) two million barrels or more of crude petroleum were produced; or (3) the Yearbook credited the county with active exploration.[11]

In multiple district counties with oil production, all members of Congress from that county were presumed to have a constituency interest in oil. This occurred mainly in Los Angeles County, California. It was felt that even if oil were not actually produced in every district in the county, it was likely, due to geographical closeness, that each district in the county contained people directly or indirectly associated with oil production.

With this operationalization of a constituency interest in oil for House districts, 117 Congressmen were coded as representing such districts in 1969. But unlike the Senate, House districts change in location and size. Given the

increased urbanization of the United States in the past two decades and the general belief that oil is produced in rural sections, we should expect that fewer House districts would have oil production in 1969 than in 1950. And if this were the case, it might help explain the reduction in the oil depletion allowance to 22 percent in 1969. However, coding House districts for 1950 shows only 115 to have oil production. This net change of two is broken down by state in Table 2-2 and gives a better idea of where the gross changes occurred. While the oil producing states of Illinois and Oklahoma lost House seats, Texas and California easily made up the difference and then some.

The overall picture is, however, one of stability. Certainly there were no fewer Congressmen from oil producing districts in 1969 than in 1950. If anything, the coding is biased in the other direction. By using 1966 data on oil production, some districts may have been included in which oil was not yet being produced in 1950. In any case the data show that reapportionment has not greatly affected the total number of Congressmen from oil producing districts. It also eliminates reapportionment as a source of explanation for the failure of the oil industry to keep the depletion allowance at 27½ percent.

If we use the figures developed here, then 27 percent of the House and 30 percent of the Senate represented districts with oil producing interests. Yet one may contend that our cutoff points for the operationalizations are, to an extent, arbitrary. Instead, suppose I only considered Congressmen and Senators from the four leading oil producing states. After all, these states do produce over 75 percent of all U.S. petroleum. Then only 8 percent of the Senate and 13 percent of the House would be considered as representing oil districts. Argument can certainly be made for the use of other cutoff points. I selected the one used here on two rational bases: (1) oil production has to be fairly large for an interest to have impact on a legislator, and (2) the size of production must be greater for the interest to affect the average Senator than the average Congressman. Any operationalization would be somewhat arbitrary. But the point here is to cast reasonable doubt on the Fenno argument at the micro-level of analysis. Depending on the distribution of the interest, it is entirely possible that a greater percentage of House members than Senators will be affected by an interest.[12] We shall explore the Fenno contention more fully at the conclusion of this chapter.

With these estimates of House and Senate districts with an oil interest, we can proceed with our analysis of whether this sizeable group of legislators with a constituency tie to the petroleum industry has an impact on the success of the oil industry in the House and Senate and in the relevant committees.

The Senate

Over the course of the depletion allowance issue the Senate has repeatedly been the more active body in considering changes in the allowance for oil. This is not

Table 2-2
Congressional Districts with Petroleum Production

State	1950 District Numbers	N	1969 District Numbers	N
Alabama	1, 2	2	1, 2	2
Alaska	–	0	AL	1
Arizona	2	1	3	1
Arkansas	4, 7	2	4	1
California	3, 6, 9-11 12-20, 22, 23	16	4, 8, 12-14, 16-26, 28-32, 34, 35	23
Colorado	2-4	3	2-4	3
Florida	6	1	9	1
Illinois	20-26	7	20-24	5
Indiana	7, 8	2	8	1
Kansas	1, 2, 4, 5	4	1, 3-6	5
Kentucky	1, 2, 7-9	5	1, 2, 5, 7	4
Lousiana	1-8	8	1-8	8
Maryland	6	1	6	1
Michigan	2-10	9	2-6, 8-10	8
Mississippi	5-7	3	3-5	3
Montana	1, 2	2	1, 2	2
Nebraska	4	1	3	1
New Mexico	2AL	2	2	1
New York	45	1	38	1
North Dakota	2AL	2	2	1
Ohio	11, 12, 16-18	5	10, 12, 16-18	5
Oklahoma	1-8	8	1-6	6
Pennsylvania	19, 28	2	23, 24	2
South Dakota	2	1	2	1
Texas	1-21	21	1-23	23
Utah	1	1	1, 2	2
Virginia	9	1	9	1
West Virginia	2, 4, 6	3	2-4	3
Wyoming	1AL	1	1AL	1
Total		115		117

AL = At Large

indicative of a more hostile attitude on the part of the Senate toward the depletion allowances. In fact, on all major legislation affecting depletion allowances, the Senate has always produced bills more favorable to the mineral industries than has the House. In 1926, when depletion for oil was set at 27½

percent, the Senate bill asked for 30 percent depletion and the House bill 25 percent.[13] In 1951 the Senate asked that the depletion allowance be extended to ten minerals not included in the House bill.[14] Finally, in 1969, when the House cut the depletion allowance for oil to 20 percent, and proportionally for other minerals, the Senate cut oil only to 23 percent, raised the percentage net limitation from 50 percent to 65 percent, and restored other minerals to their previous levels.

But floor activity on depletion has been almost exclusively confined to the Senate, which has a relatively open amendment process on the floor. House floor action on depletion has been limited, because the Rules Committee of the House traditionally has provided for a "closed rule" on all tax legislation, thus prohibiting floor amendments. (In the next chapter we shall consider the impact of the "closed rule" on the success of the industry.)

In looking at floor action on the oil depletion allowance from 1950-1969, we find ten roll call votes in the Senate dealing directly with oil depletion. In the House, only the 1969 vote on recommital of the entire bill is available. Therefore, we must rely more heavily on an analysis of the Senate to consider the full strength and impact of the constituency tie.

A few of the Senators offered all of the ten Senate amendments. John Williams of Delaware and Paul Douglas of Illinois account for seven of them. The others were offered by Hubert Humphrey, William Proxmire, and Allen Ellender. Of these, only Ellender's amendment in 1969 called for an increase in the allowance and then only to restore it to 27½ percent from the 23 percent in the Finance Committee bill.

The cutting amendments took two different forms. One called for a straight cut from 27½ percent to 15 percent. This was the change Truman requested in 1950 and again in 1951. It was introduced in an amendment by Humphrey in 1951 and by Williams in 1958. A modification of this amendment was offered by Williams in 1964 and 1969. The 1964 version called for a gradual cut over a three-year period from 27½ percent to 20 percent. In 1969 Williams' amendment was an attempt to bring the Senate bill in line with the House bill by going to 20 percent directly.

The other form of the amendment became known as the Douglas amendment. In an attempt to counter the argument that a change in depletion would discourage the little oilmen, the wildcatters who are responsible for considerable exploration, Douglas offered a graduated form of depletion allowance. Oil producers with gross incomes of one million dollars or less would still receive 27½ percent; those with gross incomes from one to five million would receive 21 percent; and those with gross incomes greater than five million would receive 15 percent. Douglas first offered this amendment in 1954, but was unable to obtain a roll call vote. It was voted on in 1958, 1959, 1960, 1962 and 1964. With the exception of the 1958 amendment offered by Proxmire, Douglas sponsored all the graduated proposals.

These ten roll call votes are presented in the following table:

Table 2-3

Senate Floor Votes on Depletion Amendments for Oil State and Non-Oil State Senators 1951-1969

		Oil State Senators		Non-Oil State Senators		Total		
		For	Against	For	Against	For	Against	Gamma
1951	Humphrey Amendment	1	25	9	55	10	80	−.607
1958	Williams Amendement	2	24	26	40	28	64	−.772
1958	Douglas Amendment	5	20	28	39	34	59	−.502
1959	Douglas Amendment	3	25	25	31	28	61	−.706
1960	Douglas Amendment	3	27	31	36	34	63	−.771
1962	Douglas Amendment	1	27	27	40	28	67	−.902
1964	Williams Amendment	1	27	34	36	35	63	−.924
1964	Douglas Amendment	1	29	34	34	35	63	−.933
1969	Ellender Amendment	25	5	7	61	32	66	+.955
1969	Williams Amendment	1	29	39	29	40	58	−.950

What this table demonstrates is the consistency of support that oil state Senators give to the depletion allowance. On only three of the ten votes did oil state Senators fail to give less than 90 percent support to maintaining the depletion allowance at 27½ percent, and the lowest percentage support was only 77 percent (20 of 26) on the 1958 Douglas amendment. Almost all the fluctuations take place among Senators from non-oil states.

In addition, the high gammas indicate that coming from an oil state was an extremely powerful criterion in determining how a Senator would vote on depletion.[15] It demonstrates a consistently strong conditional relationship.

Further evidence for the strength of the constituency relationship appears in examining the oil state defectors, those Senators from oil producing states who voted to cut the depletion allowance. Through most of the period there were only two consistent defectors. One was Douglas, and the other was John Carroll from Colorado. In the cases of both men it is difficult to say that they viewed the oil industry as a major influence in their constituencies. In a 1967 article Douglas, discussing his amendment, spoke of the relationship of the oil industry in Illinois to his position on depletion:

I must confess, however, that I also hoped that this compromise would help to split off the small operators from the huge companies and make it more possible to pass the measure. In this, I was disappointed. Although there were two or three concerns in Illinois that had gross receipts in excess of a million dollars, nearly all the small operators lined up behind the big companies and in bitter

opposition to me. They were the dominant economic interest in one Congressional district covering the southeastern section of the state. Most of them insisted that I was proposing to cut their allowance to 15 percent, although this was, of course, not the case. No explanation was effective, although finally the more knowledgeable would privately admit that while they understood that they would not be hurt, for the sake of industry solidarity they must oppose the Douglas amendment and me personally. Despite all their efforts and the oil money that came into the state in 1954 and the 1960 election campaigns, I had been able to beat off their attacks, and had even carried the oil districts.[16]

It is clear from this that Douglas realized the strength of the industry in his state. But he did not consider it as part of the constituency he need represent. Moreover, he knew that the change he was proposing would not affect most of the oil industry in his state. While the oil industry might have been an asset in an election campaign, it was not so strong in the State of Illinois that it could defeat him.[17]

The case of John Carroll is similar. As a Democrat and a liberal from an oil producing state, Carroll did not receive support from the industry. It is difficult to imagine Carroll as viewing the oil interest as a part of his supportive constituency. It is reported that oil money financed a "red smear" campaign against Carroll in 1958.[18] Further, the whole notion of a depletion allowance was in ideological opposition to Carroll's thoughts. When, during the 1958 debate over the Douglas amendment, Carroll was challenged by Senator Barrett of Wyoming on how he, Carroll, could be opposed to oil depletion and yet be the sponsor of a bill to grant depletion to oil shale, Carroll offered a most revealing response:

I am the only member of the Colorado delegation who has not sponsored it. I felt I could not be sponsoring a proposal seeking a privilege for my state in opposition to a basic principle. . . .[19]

But Carroll was realistic about the constituency problems his opposition to depletion would cause him.

I am not an expert in the oil business. But Colorado is developing great oil and gas production each year. As production grows in proportion, it may be that I shall have to change my position in the future as far as a policy on gas and oil is concerned. (Laughter)[20]

In 1962 Carroll was defeated for reelection by Peter Dominick. Dominick had served in the Colorado Legislature previous to his election and claimed that he had worked on "problems . . . connected with the mineral industry in the state" so that "most people knew of my stand in favor of developing energy sources." Dominick feels that Carroll's position on depletion "galvanized the independent oil people behind me."[21]

Of the remaining defections prior to the 1969 vote, three were one-shot affairs. These included the votes by Democratic Senators Mansfield and O'Mahoney in 1958 and Moss in 1960.[22] The one remaining defector was Senator Langer of North Dakota, whose progressive ties conflicted with support for depletion allowances. All remaining oil state Senators fit the pattern perfectly up to the 1969 vote on the Ellender amendment. The loss of five oil state Senators on that vote may indicate a limit of the industry's constituency strength. However, all but one of the five oil state Senators who voted against the Ellender amendment voted against Williams' amendment to 20 percent, and there is reason to believe that a cut in depletion to 23 percent would have at most a marginal impact on the oil industry. We shall explore this point more fully in Chapter 5.

The strength of the constituency tie is further emphasized in the floor debates on these amendments. Oil state Senators dominate the side arguing against cuts in depletion to the exclusion of nearly all others. Moreover, their statements illustrate the importance of the constituency interest in oil to their position, as the following examples demonstrate:

John Stennis (Mississippi): Without it I do not believe we would have a commercial oil well and gas well in Mississippi today.[23]

Clinton Anderson (New Mexico): I am from an oil producing state, and I have had an excellent opportunity to observe the industry's sound and stable contribution to New Mexico's economy . . . Only eight or nine days ago I was in the great San Juan oilfield in New Mexico, which is 45 miles wide and 70 miles long. . . . One of the best operators in my state is gambling $3 million on a giant project there.[24]

Mike Monroney (Oklahoma): I am sorry, but I am probably the only man in Oklahoma who does not own a nickel's worth of oil.[25]

Fred Harris: (Oklahoma): I rise to speak for my state and its interests, as does the Senator from Wisconsin when he talks about the interests of the dairy industry. . . .[26]

These are not atypical of the speeches made by oil state Senators on depletion. Nearly all speeches include some constituency reference. It is clear that oil state Senators not only are more likely to vote to protect the industry than non-oil state Senators, but they couch explanations for their votes clearly in constituency terms. As a staffer in an oil state Senator's office noted, "Oil companies are not necessarily our friends," but that being from an oil state means more than just the companies. It means the companies' employees and the general economy of the state. Thus, for example, Harris' comment to Proxmire, cited above, puts it quite clearly. It is difficult to vote against a major economic interest in your state. The oil industry has this operating in its favor. To start with, the industry knows it has the support of 25 to 30 Senators if it can remain united.

The House

While we should not expect the House to differ that much from the Senate in the strength of constituency ties of the industry to members from oil districts, the supportive data are far less extensive. In the twenty-year fight over depletion, there are only two votes that come close to being a direct vote on the depletion allowance. Both occurred during action on the 1969 bill. One was the recommital vote offered by George Bush (Rep., Texas), and the other was the vote on final passage of the bill. Neither is an accurate representation of a vote on depletion. On the former, some oil state Democrats voted against recommital on a party line basis and decided to vote against the bill on final passage instead. However, a larger number realized recommital was the only chance they had to change the bill and did not want to be on record against "tax reform" on final passage. Moreover, the bill contained sections on many aspects of tax law besides depletion. It is hard to claim that depletion was the only motivating force behind the votes of members from oil districts.

To take into account the first set of limitations, Table 2-4 uses both votes for purposes of classification. Any Congressman who voted for recommital or against final passage is classified as against the bill. Less than one half of the oil district Congressmen voted against the bill or for recommital. While this might indicate that the constituency ties of oil in the House are not as strong as in the Senate, it is more likely that this discrepancy lies in the lack of a direct vote on the depletion allowance. One should note that the gamma remains high and thus indicates a strong conditional relationship. In addition, some Congressmen from oil districts who spoke against the change in depletion either in committee or on the floor, voted for final passage. Included among these were Shriver of Kansas, Aspinall of Colorado, and Shipley of Illinois. All three were senior members in their state delegations. Shriver and Aspinall were the deans of their respective delegations and Shipley was the senior Illinois member on the Appropriations Committee. In these three state delegations twelve oil district representatives voted for the 1969 bill. While the measure of association reflects an accurate picture of constituency strength regarding oil, the distribution of oil district people on the vote itself may be misleading.

Moreover, the depletion allowance had a strong record of support in the

Table 2-4
Vote on 1969 Tax Bill

	For	Against
Oil District	68	49
Other District	287	29
	gamma = .754	

House prior to 1969. In 1950, after Truman's first attack on the allowance, thirty-three members of the House from oil districts went before the Ways and Means Committee. Congressman Herbert Meyer of Kansas claimed to be testifying on behalf of the entire state delegation. On occasion witnesses in support of depletion would be accompanied by members of Congress. A Committee member reports that one such witness, General Ernest O. Thompson of the Texas Railroad Commission, was escorted to and from the Committee hearings by Sam Rayburn. Rayburn did not introduce Thompson, and there is no record in the hearings of his presence.

Again, following the 1951 Truman proposal to cut the depletion allowance to 15 percent for oil, a swarm of Congressmen testified against the proposal. This time they were twenty-eight in number and included Toby Morris as spokesman for the entire Oklahoma delegation. By 1969, however, this was no longer the case. Less than a dozen House members went before Ways and Means to talk about depletion. While the entire Kansas and Oklahoma delegation still opposed any change, no one went to Ways and Means claiming to speak for either delegation. Thus, evidence from the floor vote supports the case of constituency ties. But by 1969 the level of activity by oil district representatives had decreased.

What is more remarkable is the absence of depletion as a subject during floor debates on tax bills. True, this is in part due to the "closed rule." However, while Douglas, Williams, and Humphrey offered floor amendments in the Senate and carried on lengthy debates with Kerr, Long, and Monroney, no mention of depletion was made in the House. In effect, it was not an issue. We will return to this subject in the next chapter, for it is indicative of the strength of the oil industry ties in the full House.

To this point the one thing that stands out is the amazing stability in the level of support for the depletion allowance among Congressmen and Senators with oil interests in their constituencies. There is one major exception to this rule, and that is "the behavior of one individual," Hale Boggs of Louisiana. The strange thing about the Boggs situation is that, until 1969, he had been considered a spokesman in the House for the oil industry. Throughout the period under study, Boggs served on the Ways and Means Committee of the House and fought to protect the depletion allowance. In 1950, he reacted to the Truman proposal in a way typical of oil district Congressmen:

... I am as anxious to plug loopholes as anyone else, but at the same time I would be reluctant to take a step which would seriously interfere with an industry which has certainly developed the southwestern part of this great country of ours.[27]

He then joined J.M. Combs of Texas in critically probing Secretary of the Treasury John Snyder and Assistant Secretaries Graham and Kirby on the proposal.

But in 1969 it was Boggs who introduced an amendment in the Ways and Means Committee to cut the depletion allowance on oil and gas to 20 percent. At that time he still claimed to have the oil industry's interest at heart. He argued "Vanik had the votes to beat me."[28] This reference was to Congressman Charles Vanik of Ohio, a member of Ways and Means in 1969 and a longtime foe of the depletion allowance.

If Boggs' statements were correct, then there is no reason to think that his position upsets the overall picture of stability just put forth. In fact, however, the available evidence proves him wrong. Four motions were offered in the Committee to change the depletion allowance on oil and gas. In order of severity in treating the oil industry they are: Sam Gibbons' motion to cut oil and gas to 16.5 percent; Boggs' motion to cut to 20 percent; Rogers Morton's motion to cut to 22 percent; and George Bush's motion to cut to 23 percent. The Committee members were scaled on these four votes and the results appear in Table 2-5. For the purpose of this table a vote for the Morton or Bush motions and against the Boggs and Gibbons motion was taken to represent pro-oil positions. Although there are errors in the scale, the results still prove most interesting. With a 25-member committee, thirteen votes are necessary to pass a motion. Considering scale position, we find that Boggs is capable of being the swing vote on the issue. True, Broyhill, Chamberlin, Mills, Byrnes, and Conable all vote for the Boggs motion. But given their position on the Morton motion, Boggs' move to 20 percent was unnecessary. He either miscalculated or had other reasons for offering his motion.

When I asked respondents whether they thought Boggs' estimate was correct or not—after all, he did not have the *post hoc* evidence that we have—I received conflicting answers. All agreed on one thing. Vanik did not have the vote in the Committee. His effort to win would be by beating the closed rule on the floor. The disagreement was over whether Vanik had the floor votes. An aide to Vanik claimed they had the votes to beat a closed rule and then to make a severe cut in depletion. But an anti-depletion Ways and Means Democrat disagreed. He felt it might have been possible under some circumstances to defeat a closed rule; however, he doubted that it would have happened if Ways and Means reported no cut in depletion. Certainly, if Vanik's office did have an accurate poll, this Committee member would have recalled it. This is confounded by the statement of a pro-depletion Ways and Means Democrat. After offering the opinion that the Committee could have produced a bill without a change in depletion, he qualified his response with, "we would have lost the closed rule on the bill." Others felt that Boggs only did it to build credibility with liberal Democrats prior to his involvement in the contest for majority leader. (Still others explained Boggs' behavior in terms of personal difficulties he was having.)

While there is disagreement over the motives for Boggs' action—some light will be brought to bear on this subject in a later chapter—oil industry representatives were united in their feelings about his action. They clearly

Table 2-5
Ways and Means Votes on Motions to Cut Depletion Allowance

	Gibbons Motion	Boggs Motion Amended	Morton Motion	Bush Motion
Landrum	+	+	+	+
Burleson	+	+	+	+
Utt	+	+	+	+
Betts	+	+	+	+
Schneebeli	+	+	+	+
Bush	+	+	+	+
Morton	+	+	+	+
Broyhill	+	−	+	+
Chamberlin	+	−	+	+
Mills	+	−	+	−
Byrnes	+	−	+	−
Conable	+	−	+	−
Boggs	+	−	−	−
Watts	+	−	−	−
Ullman	+	−	−	−
Rostenkowski	+	−	−	−
Fulton	+	−	−	−
Collier	+	−	−	−
Burke	−	−	−	−
Griffiths	−	−	−	−
Vanik	−	−	−	−
Gilbert	−	−	−	−
Corman	−	−	−	−
Green	−	−	−	−
Gibbons	−	−	−	−

Note: Italicized names are those of Democrats appointed to the Committee prior to Speaker Rayburn's death.

disagreed with Boggs' maneuver. When asked whether they felt that any segment of the industry approved of the Boggs' motion, representatives from different viewpoints within the industry offered similar comments. A representative of the American Petroleum Institute, the lobbying arm of the major oil companies, responded:

I don't think anyone approved of it. I was surprised. By that time it was apparent there would be some reduction in percentage depletion. But I don't know where he got the 20 percent figure from.

A lobbyist for the independent oil producers organization commented:

We certainly were surprised. We have no evidence that any segment of the industry supported him. Some have tried to investigate this, but haven't found anyone.

While this one vote may not be important in estimating constituency strength, it is vital to understanding the House cut in depletion to 20 percent for oil and gas in 1969.

Further, it leads us to the next step in probing the constituency related strength of the oil industry. After all, one person's vote may be unimportant on the House or Senate floor, but in a committee of twenty-five members like Ways and Means or seventeen like Senate Finance, variation of a few can be very meaningful.

The Committees—Finance and Ways and Means

While the oil industry appears to have had constituency-related ties to 25 to 30 percent of the House and Senate membership, this is hardly enough strength to win floor votes unless a large percentage of the other members are absent or can be persuaded to vote for the maintenance of the depletion allowance. Moreover, except in 1969, the tax bills that reached the House and the Senate floors contained no provisions for reduction in the allowance. If the House and Senate were to reduce the oil depletion allowance during this period, it would have been necessary for them to override decisions made by their respective tax committees. Thus, the number of members on Finance and Ways and Means with oil interests in their constituencies may provide a good indication of the reason for industry success with regard to the depletion allowance. Tables 2-6 and 2-7 present this data. Several things stand out in these tables.

First, throughout the 1950-1969 period the Senate Finance Committee has had a disproportionately large number of members from oil producing states. Given that only 30 percent of the Senators come from oil states, it is interesting that in no year was the Finance percentage that low. In fact, over the entire period it averaged nearly 40 percent. Moreover, with the exception of 1950-1952, and 1957, both political parties had higher than 30 percent of their Committee membership from oil states. (A spot check shows that for most of these years both parties in the Senate had around 30 percent of their membership from oil states. The exceptions are in the 1950-1952 period, when 24 percent of the Republican Senators and 34 percent of the Democratic Senators were from oil states.)

A second valuable point is the marked increase in oil state Senators on the

Table 2-6
Senate Finance Committee Membership

	1950	51	52	53	54	55	56	57	58	59	60	61	62	63	64	65	66	67	68	69
Number of Members	12	13	13	15	15	15	15	15	15	17	17	17	17	17	17	17	17	17	17	17
Number of Democrats	6	7	7	7	7	8	8	8	8	11	11	11	11	11	11	11	11	11	11	10
Number of Republicans	6	6	6	8	8	7	7	7	7	6	6	6	6	6	6	6	6	6	6	7
% of Democrats from Oil States	50	43	43	43	43	38	38	50	50	36	36	45	45	36	36	36	45	45	45	40
% of Republicans from Oil States	17	17	17	38	38	43	43	29	33	33	33	33	33	50	50	50	50	50	50	43
% of Total Membership from Oil States	33	31	31	40	40	40	40	40	40	35	35	41	41	41	41	41	47	47	47	41

Table 2-7
Ways and Means Committee Membership

	1950	51	52	53	54	55	56	57	58	59	60	61	62	63	64	65	66	67	68	69
Number of Members	25	25	25	25	25	25	25	25	25	25	25	25	25	25	25	25	25	25	25	25
Number of Democrats	15	15	15	10	10	10	15	15	15	15	15	15	15	15	15	17	17	15	15	15
Number of Republicans	10	10	10	15	15	10	10	10	10	10	10	10	10	10	10	8	8	10	10	10
% of Democrats from Oil States	20	27	27	20	20	20	20	20	20	27	27	20	20	20	20	18	18	13	13	20
% of Republicans from Oil States	30	30	30	20	20	10	10	10	10	20	20	20	20	20	20	25	25	30	30	20
% of Total Membership from Oil States	24	28	28	20	20	16	16	16	16	24	24	20	20	20	20	20	20	20	20	20

Committee after the 1952 elections. This neatly follows the Truman attack on the depletion allowance. One may claim that this increase should have occurred following the 1950 elections. However, Finance had only one vacancy in 1951, and that was a Democratic one. Table 2-6 shows that the increase in oil state Senators on the Committee was due to a change in Republican membership. The Democrats already had over 40 percent of their Finance members from oil states.

Third, we find the Ways and Means Committee tends to underrepresent oil interests. While there is some variability in oil representation, it tends to stay around the 20 percent figure (or five members on the 25-member Committee). However, 27 percent of the House members represent oil districts. One would expect that proportionally the committee would average seven oil district Congressmen in its membership. (The reasons for this phenomenon will be discussed in the next chapter when we look at the recruitment process for Ways and Means and Finance.)

What jumps out at us at this point is a prime reason why the oil industry did better in tax matters in the Senate than in the House. The data in Tables 2-6 and 2-7 show that the Finance Committee consistently has a higher percentage of members from oil constituencies than does the Ways and Means Committee. Given the complexity of tax legislation, the success of these two committees on the floor of their respective bodies, and their importance when evaluated by members, it is not surprising that this difference in committee membership is congruent with industry success.[29] As one industry representative put it in commenting on where the industry tries to exert its influence:

The committees which write the legislation are the biggest point of concentration. Most things after that are just a salvage job. . . . When you really get down to it, it's how you influence the two committees.

The interview data substantiate the findings on oil district membership on Ways and Means and Finance. When asked why the oil industry did better on tax legislation in the Senate than in the House, many respondents mentioned the make-up of the committees among their reasons. Again the same response came from differing points of view:

A liberal Democratic Senator: That Committee (Finance) has been loaded in oil for years.

A Ways and Means Republican: The Finance Committee is composed of Senators from mineral and farming states whose main purpose is representing their constituencies.

An Oil Lobbyist: You had people on Finance that represented oil states . . . Long and Kerr before him and others from oil producing states were in positions of importance.

Some preferred to claim that the Senate, in general, had a higher percentage of members from oil constituencies than the House. The data show that this difference is marginal when compared to the difference in committee membership.

While the membership data on the committees are revealing in estimating differences in the industry's success between the House and the Senate, it should still be noted that industry strength never reaches 50 percent at any of the four decision making points so far examined. Of course, it may not be necessary to have strength greater than 50 percent. Many who do not have oil interests in their constituencies may not care about the depletion allowance enough to build coalitions in an effort to cut it. But this has not always been the case with the oil depletion allowance. Certainly, when Truman put his support behind a cut to 15 percent, the industry was very much aware of the problem it faced. It is interesting to see how the industry bargained for additional support. It is a classic example of the use of the "simple logroll," the most obvious form of negotiated bargaining. As Lewis Froman has noted:

By its very nature, then, simple logrolling is explicit. To engage in this type of bargain each of the parties has to know what he wants and what he is willing to support in return for support from others.[30]

This is precisely what transpired in 1950 and 1951 following Truman's request for a cut to 15 percent depletion for oil and gas. When Ways and Means met on these two revenue bills, rather than cutting oil and gas, they proceeded to add more materials to the list of depletable minerals. The 1950 attempt to add minerals did not succeed due to the request for increases in taxes to pay for the Korean War. In 1951, however, there was no holding back, and some thirty-one new substances were added to the list of those receiving depletion. Thus, early in the 1951 Ways and Means hearings, Charles E. Brady from Salisbury, North Carolina, testified before the Committee. Brady was representing the National Sand and Gravel Association, and requested depletion allowances for those substances. Salisbury, North Carolina, just happened to be in the district represented by Robert L. "Muley" Doughton, then Chairman of Ways and Means.[31] Later, in the hearings, Richard Butler Carothers of the H.C. Spinks Clay Company of Paris, Tennessee, testified on behalf of the clay industry. He requested that ball clay be kept at 15 percent depletion rates. Paris, Tennessee, was also represented on the Committee. Jere Cooper, next in line for the chairmanship, came from the district that included Paris.[32] Other witnesses had similar ties. True, the relationships were not always so neat, but nevertheless, clear constituency ties existed. For example, substantial quantities of limestone, another material added in 1951, was being quarried in districts represented by Wilbur Mills, Noble Gregory, Sidney Camp, Burr Harrison, Richard Simpson, Noah Mason, and Thomas Martin. All were members of Ways and Means in

1951; none had oil production in their districts. With the exception of a handful of members, everyone who sat on Ways and Means in 1951 saw a resource produced in his district being added to those receiving a depletion allowance.

Even Herman Eberharter, a Pennsylvania Democrat and the only Committee member who ever publicly denounced depletion allowance, became noticeably less hostile to industry witnesses once the coal and perlite industries, both important in his district, had submitted requests for increase and inclusion, respectively, in depletion allowances. During the floor debate on the bill, Eberharter paused only briefly in his defense of the bill to comment on depletion:

I do not approve of some of the depletion allowance, at least, to the extent granted under the bill.[33]

He did not elaborate, but one doubts he objected to the increase for coal.

The Senate Finance Committee members went through a similar process in 1951. Not only did they grant the allowances or increases to the twenty-one substances the House had acted on, but they added ten resources. Naturally, by the time the bill had reached this stage every Finance Committee member with any mineral production in his state was covered by the change. The logroll had become tremendous.

The Senate floor debates provide assurance that this logrolling was not an accident, but was, in fact, an explicit bargaining maneuver. Instead of defending depletion on substances produced in their home state, Senators deliberately chose to speak for those from another state. Matthew Neely of West Virginia, in a speech covering three pages of *The Congressional Record*, did not mention the increase in depletion allowances for coal from 5 percent to 10 percent, but spoke only about oil.[34] Neely was followed not long after by Pat McCarran of Nevada, who defended the allowance for oyster shells with the following preface:

Any of the opponents of depletion allowance for mines that desire to be additionally acid in their comments on this measure can hardly accuse a Nevada spokesman of being personally interested in oyster shells.[35]

At an early point Humphrey offered his own moral analysis, but put the situation in clear perspective:

Let me say that the contention that the new minerals which have been added in this version of the bill are competitive with some which are enjoying the privilege is a contention which reminds me somewhat of the allegation that after one tells a lie the only way to cover it up is to tell another one, and thus keep getting deeper and deeper into the complication and the trouble. So, in the present case, apparently it is felt that the one way to proceed is to add others.[36]

One should not get the impression that 1951 was the only year logrolling was used to protect the oil depletion allowance. In the 1958 debate on the Douglas amendment John Carroll, speaking in favor of it, noted:

If the amendment fails . . . perhaps the time has come, under the doctrine of equation, to take another look at oil shale and perhaps give proper depletion allowance treatment to oil shale.[37]

This extensive logrolling is more than just an isolated example. It represents a situation in which private interests benefit from the "socialization of conflict." What the industry and its supporters in Congress foresaw was that the oil depletion allowance did not have enough constituency related support to survive an attack from an active oil depletion opponent. The strategy therefore became one of redrawing the cleavage lines by expanding the conflict. However, in this case, expansion of conflict does not have Schattschneider's predicted effect of weakening private interests.[38] Instead it improved their position. The logroll is a general form of such conflict expansion for which depletion has been a specific example. It allows for more private interests to be aggregated so that a winning coalition can be maintained. Further, these findings are not inconsistent with Schattschneider's general point about socialization of conflict. Rather the findings refine his statement by placing some limits on them.

Oil Depletion and the Treasury Department

The Congress is not the only decision making point in the government to which the oil industry has constituency ties and other forms of access. The industry also has ties to the Executive agencies. In examining the depletion allowance, we would like to know how important these constituency ties between the oil industry and the Treasury are to the industry's success. Such ties are much more difficult to substantiate than those with the Congress. Clearly there have been differences in the success of the industry congruent with the leadership of the Treasury Department.

In the years from 1950-1969 there have been two cases where the oil industry has had a direct tie with the Secretary of the Treasury. Both of President Eisenhower's Treasury Secretaries were directly involved with the industry both before and after their terms at Treasury. George Humphrey came to Treasury from M.A. Hanna and Company, whose interests included sizeable investments in oil. When Humphrey left the Treasury in 1957, he returned to his position with Hanna. Robert B. Anderson, who succeeded Humphrey, had an extensive background with the oil industry. Among the posts he held before going to the Treasury was as a director of both the American Petroleum Institute and the Independent Petroleum Institute of America, the two leading lobbying arms for the oil industry. In 1951, as President of the Texas Midcontinent Oil and Gas

Association, he had testified in defense of the depletion allowance.[39] In 1957, at his confirmation hearing before the Senate Finance Committee, Anderson was questioned by Senator Albert Gore regarding the 1951 statement. Anderson responded, "I will approach this as objectively as I possibly can and will be pleased to cooperate with Congress . . . in a review of depletion allowances."[40] But as with Secretary Humphrey, no review occurred.

In 1970 a *Washington Post* article reported that Anderson entered into an agreement with oil company officials on June 14, 1957, prior to his becoming Secretary of the Treasury. Under the arrangement Anderson would receive two payments totalling $900,000. Half was to be paid over a four and a half year period while the remainder was contingent on revenue produced by oil wells on certain properties. Anderson called the arrangement "ordinary, normal," but denied it influenced his position on the oil import quota system. Whether he was or was not influenced is not the point here, rather it was just further evidence of the close tie between Anderson and the industry even while he was at Treasury.[41]

It has been only during the years of the Humphrey-Anderson reign at Treasury that there was no attempt on the part of the Department to investigate the depletion allowance. Under every other Secretary of the Treasury during the 1950-1969 period, some work on lowering of the allowances was undertaken.

Industry representatives freely admitted that it was "more comfortable" to have John Connally as Secretary of the Treasury.[42] They felt he fit into a group with Anderson and Humphrey. Yet, as one industry spokesman commented, the fact that a large group of higher civil servants holds over from one Administration to another dampened this feeling of comfort considerably. He quickly qualified this, however: "but at least you feel your position is being enunciated at high levels . . . and correctly."

It might be added that this confidence is not unfounded. Replying to questions about further tax reform, Connally told the Senate Finance Committee in early 1972, "I don't consider the (oil) depletion allowance a loophole."[43]

The general opinion of the oil lobbyists was that on occasions when their position was not being voiced strongly at Treasury, as was the case from 1961-1968 under Treasury Secretaries Dillon, Barr, and Fowler, the Department was to be feared. It was the one point in the process where the industry felt particularly weak. After all, one of Treasury's duties is to locate sources of revenue. Depletion allowances protected a source of revenue. Thus, when the Department was only able to get small changes in depletion through Congress during the early 1960s, such as disallowing property grouping in computing gross income,[44] it began to look for new ways of limiting the revenue loss from the depletion allowance. The instrument employed for this purpose was a new set of rules by the Internal Revenue Service setting the cutoff point at which oil and gas prices were set. These rule changes required no approval by Congress.[45] However, following a series of meetings between Treasury officials and oil

industry representatives, they settled the disagreement, and the original pricing system was for all purposes maintained.

I do not wish to give the impression that Treasury and the oil industry are at each other's throats. True, when someone like Stanley Surrey, an academician with strong feelings about tax reform, is Assistant Secretary for Taxation (1961-1968), the industry does face some problems.[46] But the industry's problems with the Department are not really as grave as they might seem. The frequency of private meetings between the oil industry and the Department is indicative of this.[47] In the year from August 1968 to July 1969, at least eight formal meetings took place to consider various aspects of the depletion allowances. One former Treasury official explained the frequency of these meetings as an "information" exchange. He claimed that industry representatives did not want to be surprised by proposals Treasury made to Congress, and that Treasury needed to prepare for the industry's counter arguments. The meetings were just a form of dress rehearsal to insure that nobody looked bad and to allow both sides to develop and modify their positions.

Another Treasury official was considerably more cynical about this high level of interaction, which he claimed existed between Treasury and many industrial groups:

The reason why there is such a close working relationship between industry and Treasury is that they've been in bed with each other for decades. . . . It's the same all over. For example, the law firm of Covington and Burling has an endowed chair in the Solicitor General's Office. By this I mean that young lawyers in that firm will go to work at the Department for two years and then return to Covington and Burling only to be replaced by another young lawyer. Don't get me wrong, it's not nefarious. It's personal. They probably have fifteen or sixteen members of the firm who have direct connection with Justice.

When asked whether there were similar cases at Treasury, the respondent said that most of the sixteen or so young tax lawyers who worked for Surrey and later for Ed Cohen and Jack Nolan—Assistant and Deputy Assistant Secretaries for Taxation in the Nixon Administration—were now with law firms representing corporate clients. Two of the group now work for McClure and Trotter, the lobbying arm of Mobil Oil. Nolan has returned to Miller and Chevalier, the law firm that provides the tax counsel, David W. Richmond, for the American Petroleum Institute (this is in no way meant to be critical of Nolan, who receives high praise even from the leaders of public interest tax groups, but merely to provide an example of the practices). But again, this respondent emphasized that this was a normal course:

That's the characteristic job pattern. Their individual knowledge and access is worth something here and not in private practice in Peoria or wherever.

Interestingly, this individual, upon leaving Treasury, was replaced by Burke Willsey from Miller and Chevalier. Willsey had been an assistant to both Richmond and the counsel to the American Mining Congress.

Stanley Surrey is uncharacteristic of the Treasury pattern. After his service at Treasury, he returned to Harvard Law School. But individuals like Surrey can be constrained in their zeal for tax reform. Manley writes about the extremely difficult time Surrey was given by the Senate Finance Committee during his confirmation hearing. Specifically, he refers to Senator Clinton Anderson of New Mexico forcing Surrey to admit that he had "an open mind on the question of depletion."[48] While sources close to Surrey claim this hearing was no more a real conflict than a professional wrestling match is a fight, others hold to the belief that Finance members wanted to be sure that Surrey's authority was limited. One respondent claimed that a *New York Times* reporter covering the confirmation hearings thought it was the real thing and the most disheartening day in Surrey's career. And in discussion with people familiar with tax politics, Surrey's name regularly brought forth comment that his major weakness was a lack of a "Washington legal–political" experience and know-how.

One wonders, after listening to the comments of both industry representatives and Treasury officials, if they are referring to the same relationship. How vulnerable can the industry be at other decision points if this picture of Treasury is at all accurate?

To some extent, the latter part of this discussion about the relationship between Treasury and the oil industry falls outside traditional notices of constituency. Yet, the symbiotic relationship comes very close to the often cited situation in regulatory politics of the regulator being almost indistinguishable from the regulated.[49] The main difference is that a law firm acts as the surrogate for the industry. This being the case, the relationship is comparable to the constituency relationship discussed earlier.

Oil Depletion–The President

Just as Congressmen and Senators have constituencies composed of various interests, so also does the President have a constituency. His is far broader than others. It might be considered fair to claim that all Presidents of the United States to some extent have the oil interest as part of their constituency. This, however, would be misleading. When an interest group actively campaigns for a candidate's opponent, it is hard to believe that the candidate, if elected, will favor that interest with his support. Just as Senator Douglas could overlook the oil industry in Illinois, one expects Presidential candidates, who were opposed by the industry, might oppose the industry once elected.

Some interest groups cover their bets by contributing to the candidates of both major parties.[50] Others, however, almost exclusively contribute to only one party. In this latter case the constituency tie is not established if the other party's candidate wins. In fact, one of the main reasons for contributing to only one party may be that the contributing group feels that the candidate of the other will oppose the group's position on certain issues, regardless of whether the group supports him or not. With one exception in the period under study,

the oil industry behaved in this fashion. From data available on campaign contributions in Presidential elections from 1968, we find that the oil industry contributed almost exclusively to Republican candidates. Table 2-8 illustrates the point. With the exception of 1964, the industry gave better than 90 percent of its contributions to Republicans. The 1964 exception is usually explained in terms of Lyndon Johnson's close ties to the oil industry and the low probability of a Goldwater victory.[51] Still, A.P.I. contributions ran 2:1 against the Democrats.

While no exact figures are available on the industry's contributions in the 1948 election, it is believed that the Dixiecrat ticket of Thurmond-Wright received substantial support from oil.[52]

Campaign contributions may just be symptomatic of a constituency tie. However, evidence does exist to support the linkage that has been drawn. When Treasury and IRS decided to make rule changes on depletion in 1968, oil trade publications were enraged that Lyndon Johnson had violated the *quid pro quo*. *U.S. Oil Week* accused Johnson of taking this action after he had "supped at oil's table for three years."[53] *Parade Magazine*, reporting the story for a mass audience, provided a more graphic description:

What oil company executives are calling the President in private is not fit to print. . . . That Lyndon Johnson, one of their own Texas boys to whose

Table 2-8
Contributions in Presidential Election Years by the Oil Industry

	Republican	Democratic
1972-Senior Oil Company Officials and Stockholders to Nixon Campaign[a]	$4,981,840	
1968-American Petroleum Institute[b]	$ 429,366	$30,606
1968-Contributors of $10,000 or more in the Oil Industry	$ 739,485	$45,000
1968-Independent Petroleum Association	$ 90,000	$ 2,500
1964-American Petroleum Institute	$ 48,310	$24,000
1960-American Petroleum Institute	$ 113,700	$ 6,000
1956-American Petroleum Institute	$ 171,750	$ —
*1952-American Petroleum Institute[c]	42	1

*Figures are for numbers giving $500 or more

[a]This figure was computed by Congressman Les Aspin (Dem., Wis.) based on contributors data available from Common Cause and the General Accounting Office. No estimate was made of oil industry contributions to McGovern.

[b]Herbert Alexander, *Financing the 1968 Election* (Lexington, Mass.: Lexington Books, D.C. Heath, 1971), pp. 327, 346, and 184.

[c]Heard, op.cit., p. 101.

campaigns they contributed a small fortune, that "ole Lyndon" should now try to slip the knife—for them such behavior constitutes unpardonable perfidy and at best a persecuted sense of financial patriotism.[54]

The fact that Presidential candidates see the oil industry as part of their constituency is clearly attested to in the 1968 campaign. In nearly every speech that Nixon made in Texas he mentioned the depletion allowance and announced his support of it. On November 1 and 2, on a last swing through the state, he spoke at Lubbock, Fort Worth, Austin, El Paso, and San Antonio. A major point in his speech in each case was the depletion allowance.[55] After the election it was reported that Hubert Humphrey, immediately following the Democratic National Convention, had tried to obtain financial help from the oil industry. In a Houston meeting Humphrey supporters were asked by industry representatives what the Vice President intended to do about the depletion allowance. A Humphrey supporter claimed "We told them the Vice President would not promise a thing. . . . The oil people wouldn't give us a dime."[56]

After the election the oil industry clearly expected that Nixon would protect the depletion allowance. The *Wall Street Journal* cited the industry's optimism:

'One of the things we knew before the election,' says W.W. Keelen, chairman of Phillips Petroleum Co., 'was that Mr. Nixon has always been for protecting the depletion allowance even though there will be a fight over it in Congress next year.' Asserts A.W. Farkington, president of Continental Oil Co.: 'Mr. Nixon's record is clear in his support of the depletion allowance, and it will survive the 91st Congress.'[57]

Available evidence indicates that the industry's optimism was well founded. In spite of the cut that occurred in the 1969 bill, President Nixon was one of the last holdouts in favor of maintaining 27½ percent depletion. Drew Pearson and Jack Anderson reported that Nixon gave Ed Cohen only "one inflexible guideline" for preparing the Administration's tax reform package—"that there be no change in the oil depletion allowance."[58] This is substantiated by a former Treasury official's claims that the Administration "dragged its feet." He reports that, following House passage of the 1969 Tax Reform Bill, Assistant Secretary of the Treasury Ed Cohen went to meet with the President at San Clemente. Cohen tried to persuade the President to come out with his own plan for changing the allowance. The plan would take the amount of money a company used for exploration into account for computing depletion. Nixon refused to support it or any change in depletion. Later, in a meeting between Treasury and industry representatives, Cohen quoted Nixon as saying, "I made a campaign promise and I intend to keep it." At that same meeting, an industry official voiced his appreciation of Nixon's pledge and for the access the industry had to the Treasury. He claimed, "If a Democrat were in power, we would not have this power."

Evans and Novak, in their book on Nixon, provide a different report of the San Clemente meeting. Their source, named by one of my respondents as being Charles Walker, then Under Secretary of the Treasury, claims that Nixon inquired about the likelihood of a cut in the oil depletion allowance and was informed that it was probable, although the final figure would be higher than the House-passed 20 percent. To this he is supposed to have responded, "Well, we might as well go with it. Let's just accept it and go on from there." The Treasury plan was never presented as an alternative. Even if this latter story is the correct one, it still shows Nixon's reluctance to go along with a change in depletion.[59]

Summary: Constituency and Depletion

There are several points that can be gathered from this examination of the oil industry's ties to the decision making process on the depletion allowance. First, the basic position set forth by David Truman is reinforced by these findings:

The separation of powers, especially between the Legislature and the Executive, and the accompanying system of checks and balances mean that effective access to one part of the government, such as the Congress, does not assure access to another, such as the Presidency. For the effective constituencies of the Executive and the members of the Legislature are not necessarily the same, even when both are represented by men nominally of the same party. These constituencies are different. . . .[60]

The strength and number of constituency ties the industry has at various decision making points makes a difference at the level of success attained. This is true both within and among the various decision making points. It shows in the differential treatment of the industry by the House and Senate Committees; the activity of the Treasury Department under different leaders; and variation in Presidential initiative to reform depletion. Variation cannot be found everywhere, but this may be due to a consistency in the level of active constituency ties—as with the Senate Finance Committee. As we proceed, refinements will be made on this conclusion.

Second, the greatest concern for the industry on the depletion allowance has been the Treasury Department and the President. Compared to the Congress, the composition of decision makers at these levels has been far less stable. This is in part due to the number of decision makers at these points. As long as oil is produced as extensively as it is now, the industry can always count on 30 percent of the Congress having constituency interest in oil. However, a President or a Secretary of the Treasury and a couple of Assistant Secretaries provide the opportunity for much volatility at Executive decision making points. Even the backlog of senior civil servants does not insure the industry a stable level of treatment at Treasury. In a speech to the National Tax Association, Thomas Field, a former Treasury official and Executive Director of Taxation with

Representation, a public interest tax organization, claimed civil servants were important in formulating the policy, but he warned:

It is easy . . . to overestimate the role of the Treasury Department in the formulation of tax policy. Like any Executive agency, the Treasury Department is subject to political considerations which set limits on its reform initiatives. The privileged status of the oil depletion deduction under Presidents Johnson and Nixon is a case in point.[61]

Given the often expressed belief that tax legislation is one area where Congressional influence on policy is a significant challenge to Executive decision making, the concern of the oil industry with the Executive may be surprising. But as Manley has noted, Congress is not prone to overriding Treasury proposals.[62]

Third, at each decision making point constituency ties between the industry and the decision makers are strongly associated with the position taken by decision makers on the oil depletion allowance. Whether the tie exists because the decision maker has production in his constituency, or because the industry contributes heavily to his campaign (either due to his position or to influence his position), or because he is directly connected to the industry is not particularly of concern to us here. Our concern is that, whatever the tie, decision makers with oil constituencies overwhelmingly tend to favor the maintenance of the depletion allowance and that their explanations rest to a significant extent on constituency support.

Finally, the examination of constituency ties has enabled us to understand the actions of the industry when the depletion allowance has been attacked. Lacking majority constituency ties on the Ways and Means Committee in the early 1950s, the industry resorted to a controlled expansion of the conflict through the use of logroll. Whether the threats of 1950 and 1951 to cut depletion really placed this provision in danger is questionable. But the industry's reaction to the threat was to bury the issue as quickly and as completely as possible.

Our attention now turns to the importance of the constituency relationship on water pollution legislation.

Constituency Support and Water Pollution Legislation

The House and the Senate

When examining the constituency ties of the petroleum industry in the House and Senate on water pollution legislation, obviously there is no change in the number of ties from that on depletion. What does change is the effectiveness of those ties. On depletion, legislators with oil constituencies had a strong tendency

to vote the industry's position even on the floor. The same does not hold on water pollution legislation. A water pollution bill that reaches the House or Senate floor rarely attracts opposition on a recorded vote. Amendments are few and most involve project authorization levels, not establishment of enforcement procedures and fines. In the 1965-1970 period, when the oil industry took special interest in water pollution bills, opposition on the floor to water pollution bills from oil district representatives was not observable from roll call votes (see Table 2-9). This is in marked contrast to votes on depletion.

Only the 1965 Federal Water Pollution Control Act had opposition in the form of floor votes and substantive amendments. But votes on the two Senate amendments designed to limit the power of the Secretary of Health, Education, and Welfare to promulgate Federal water quality standards split on party lines, not constituency lines. Both were opposed by oil state Democrats like Long, Ellender, Harris, and Monroney. True, 4 of 9 opponents on final passage (8 votes and 1 pair) were from oil states, but so were 25 Senators who supported final passage.

Moreover, none of the bills faced a recommital vote in the House.[63] It would be wrong, however, to conclude from this that the constituency ties of the oil industry to House and Senate members were ineffective. Rather, the oil industry sees the House and Senate floors as the wrong place to have impact on legislation. As one spokesman put it, "Once you get to the floor, public support on the side opposed to industry is too strong." Instead, the industry chooses other decision points at which to assert its influence. Points of lower visibility are used. Indicative of this is the fact that while water pollution bills pass the Senate in 1967 and both the House and Senate in 1968 and 1969, no bill reaches the President for signing until 1970.

The lack of constituency related support for the oil industry on the water pollution roll calls does not mean that debates were totally devoid of comment from oil district members. Russell Long challenged sections of the 1968 bill with the following comment to Senator Jennings Randolph:

Table 2-9
Floor Votes on Water Pollution Legislation

Year	Bill	Senate	House
1965	S 4	68-8	396-0
1966	S 2947 HR 16076	90-0	313-0
1967	S 2760	voice	no action
1968	S 3206	1st action voice	277-0
		2nd action voice	2nd action unanimous consent
1969	HR 4148	86-0	392-1

As the Senator knows, this measure involves oil pollution, and Louisiana is one of the great oil producing states of the Nation. . . .

Can you picture the economic chaos resulting from the shutdown of thousands of oil wells and several dozen small inland refineries?[64]

But, unlike the Senate debates on depletion, Long's speech was the exception and not the rule. Few stood to defend the industry.

The oil industry did not totally ignore floor action. But only when that action had low visibility did the industry attempt to influence decision making. For example, S.3206, the 1968 bill, passed the House and Senate on two different occasions. Each time the bill had considerably different sections regarding spillage of oil from vessels, offshore facilities, and onshore facilities. At 12:55 P.M. on October 14, the House accepted all the sections of the second Senate action except the amendments covering liability for onshore and offshore installations. *CQ* reported heavy lobbying on behalf of the industry on this final day of the session.[65] But the vital point is that the bill, which still contained a section covering spillage from vessels, did not reach the Senate before adjournment at 2:17 that afternoon. Normally this process takes no longer than a few minutes. Interviewees generally intimated that the industry, with the cooperation of Russell Long, the Senate Majority Whip, had something to do with this delay. However, industry officials were unwilling to comment on this.

The Committees

As with depletion legislation, if we really desire to estimate the effectiveness of constituency ties, we must examine the composition of the committees and, in this case, the subcommittees that produced the legislation. In the House major water pollution legislation was handled by either the Public Works Committee or its Rivers and Harbors Subcommittee during the 1955-1969 period. Even in 1968, when the Rivers and Harbors Subcommittee took charge of the bill,[66] there existed little difference between it and the full Committee in membership composition. Twenty-one of the 34 members of the full Committee had served on Rivers and Harbors. And throughout this period John Blatnik of Minnesota, either through his formal position as Chairman of Rivers and Harbors or through informal action chaired the full Committee's hearings on water pollution legislation. Both Public Works Chairmen, Charles Buckley and George Fallon, stepped aside and let Blatnik "run the show" when the legislation was handled by the full Committee. This fits with the general pattern of subcommittee autonomy operating in the House Public Works Committee.[67]

Given, in addition, that "Public Works is a successful committee—a committee which typically elicits a favorable House response,"[68] we should not be surprised by the low level of floor activity on water pollution bills. Most of the conflicts are settled in the Committee with the exception of those existing

between the House Committee and the Senate Committee. Thus, the oil industry's impact on the House bill, as with depletion, must be made at the committee level.

Until 1963, the Rivers and Harbors Subcommittee of the Senate Public Works Committee considered Senate water pollution legislation. On January 1, 1963 Senator Robert Kerr of Oklahoma died. He had served as Chairman of the Rivers and Harbors Subcommittee and was scheduled to assume chairmanship of the full Committee because Dennis Chavez, the previous Chairman of Public Works, had died in late 1962. The chairmanship of Public Works thus passed to Pat McNamara of Michigan. Compared to Kerr and Chavez, who opposed strong federal control of pollution, McNamara was favorable to new enforcement legislation. McNamara assumed control of Rivers and Harbors but created the Special Subcommittee on Air and Water Pollution. He gave the chairmanship of this new subcommittee to Edmund S. Muskie of Maine. Muskie had previous involvement in the subject matter both as Governor of Maine and as a member of Rivers and Harbors.[69] Since 1964 most major Senate water pollution legislation has come from the Muskie Subcommittee.

Thus, in examining the constituency ties of the oil industry to Congressional decision making on water pollution, we are especially interested in the composition of the Public Works Committee in the House (and in 1967 and 1968 its Subcommittee on Rivers and Harbors) and the Subcommittee on Rivers and Harbors and, after 1963, the Special Subcommittee on Air and Water Pollution in the Senate. Tables 2-10 and 2-11 provide the data on the percentage of seats on these committees filled by representatives from oil producing districts during the 1955-1969 period.

The reason for selecting 1955 as a starting point is that prior to 1955 Congressional debate centered on establishing the existence of a pollution problem. While attempts to pass water pollution legislation began toward the end of World War II, and in fact a major bill was passed in 1948, it was not until 1955 that attempts to include significant federal enforcement in legislation were considered.[70] The Water Pollution Control Act (S.418) produced little conflict. The fact that its co-sponsors were Alben Barkley of Kentucky, a Democrat, and Robert A. Taft of Ohio, a conservative Republican, and that most of the authorization was designed to benefit the Ohio River Valley, gives some indication of the broad range of its ideological support and of its pork barrel content. Only when requested by states adversely affected by interstate pollution could the Public Health Service begin limited enforcement procedures. The enforcement sections were so weak that the only opposition to the bill, when it reached the floor of the House, came from those who viewed it as inadequate. Karl Mundt, then a Congressman from South Dakota, speaking in opposition to S.418 suggested, "when the polluters of America favor that bill, there must be good cause to question the efficacy and the adequacy of such a legislative proposal."[71]

Table 2-10
Senate Public Works Subcommittee on Rivers and Harbors (1955-1963) and Air and Water Pollution (1964-1969)

	1955	56	57	58	59	60	61	62	63	64	65	66	67	68	69
Number of Members	11	11	11	11	13	13	15	15	17	9	9	9	11	11	10
Number of Democrats	6	6	6	6	8	8	9	9	12	6	6	6	7	7	6
Number of Republicans	5	5	5	5	5	5	6	6	5	3	3	3	4	4	4
% Democrats from Oil States	17	17	33	16	13	13	22	22	17	17	33	17	14	14	17
% Republicans from Oil States	20	20	20	20	0	0	0	0	20	33	67	67	25	25	25
% Committee from Oil States	18	18	27	18	8	8	13	13	18	22	44	33	18	18	18

Table 2-11
House Public Works Committee

	1955	56	57	58	59	60	61	62	63	64	65	66	67	68	69
Number of Members	34	34	34	34	34	34	33	34	34	32	34	34	35	34	35
Number of Democrats	19	19	19	19	22	22	20	20	20	18	23	23	20	19	17
Number of Republicans	15	15	15	15	12	12	13	14	14	14	11	11	15	15	15
% of Democrats from Oil Districts	42	42	37	37	32	32	30	30	35	33	35	35	40	42	47
% of Republicans from Oil Districts	20	20	13	13	8	8	23	21	14	14	9	9	13	13	13
% of Total Membership from Oil Districts	32	32	26	26	24	24	27	26	25	26	26	26	29	29	29

This does not mean we shall ignore legislation earlier than the 1955 bill. The symbolic quality of the S.418 bill will be a focus of attention at a later point. But for evaluating the strength of constituency ties of the oil industry in the committees, we shall begin with the first major attempt at federal enforcement in 1955.

In examining Table 2-10 one is immediately struck that, with the exception of the 89th Congress, the oil industry is underrepresented on the Senate Public Works Subcommittee dealing with water pollution legislation. In the 1967-1969 period, when the Muskie Subcommittee reported legislation designed specifically to deal with oil pollution, only two members of the Subcommittee, Murphy and Montoya, came from oil producing states. Throughout the 1955-1969 period, the industry lacked sufficient constituency ties to have a serious impact on the Subcommittee action.

The House Public Works Committee presents a very different case. While membership on the Committee closely approximates the proportion of oil district and non-oil district Congressmen in the full House, there is a marked discrepancy between the Democrats and Republicans on their contribution to this balanced ratio. The Democrats consistently overrepresent oil districts on the Committee, and the Republicans, just as consistently, underrepresent them. From this one might hypothesize that this merely reflects a higher proportion of oil district Democrats in the House. But this is not the case. In 1969, for example, 65 of 242 Democrats (27 percent) and 54 of 189 Republicans (29 percent) represented oil districts.

Given the highly partisan nature of the House Public Works Committee,[72] the fact that the Democrats controlled the Committee during the time period, and the general tendency of Republicans to be more favorable to the industry on pollution matters, the high percentage of Democratic Congressmen from oil districts is very significant. Further, the percentage of Democrats from oil districts reached a peak in the 1967-1969 period.

It is not surprising, given this data, that the House Committee produced weaker water pollution legislation than the Senate Committee from 1964-1969. Chart 2-1 briefly lists the major differences between the enforcement sections of House and Senate bills during this period.

Oil industry officials hesitated to comment on committee composition in explaining their success (or lack of it). But they consistently expressed the belief that the House Committee was more "reasonable" in considering the industry position. When presented with the data, however, they were willing to admit that constituency differences between the two committees had some impact. One official was quick to qualify by saying:

But people from oil constituencies are by no means a majority. I doubt if there's any committee with more than 30 percent from oil constituencies. . . . Sure it makes some difference in getting our position across, but it's just of marginal importance.

Chart 2-1

Enforcement Provisions in Senate and House Water Pollution Bills 1965-1969

	Senate Bill	House Bill
1965	Authorizes Secretary of Health, Education, and Welfare to establish Water Quality Standards if state fails to set standards in a reasonable amount of time which the Secretary considers accurate.	Drops Senate provision giving Secretary authority to set standards. Instead allows Secretary to withhold federal pollution grants from states which fail, within 90 days of bill, to file notice of intention to set standards.

1966	Amends Oil Pollution Act of 1924.
	1. Interior Secretary to regulate the discharge from vessels.
	2. Fines for violation of Act:
	  a. for individual
	    –up to $2500 and 1 year in prison
	  b. for vessels and shore installations:
	    –up to $10,000 fine
		No stiffening of oil pollution penalties.

1967	Repeal of 1924 Act Authorize 1. Interior Secretary to have discharged oil removed from water and shoreline. 2. Sets fines at same level as in 1966 bill. 3. Makes vessels and installations liable for oil discharge whether willful or accidental.	No action on Senate Bill.

1968 (Final Action)	Sets liability limits for discharge
	1. vessels
	2. onshore installations
	3. offshore installations
	  –$5 million
		Sets liability limits for willful or negligent discharge.
		1. vessels
		  –$5 million or $67/gross ton
		2. onshore installations
		  –spill prima facie evidence of liability

1969	Sets liability limits for discharge other than willful or negligent
	1. vessels
	  –$14 million or $125/gross ton
	2. onshore installations
	  –$8 million or $125/ton
	3. offshore installations
	  –$8 million or $125/ton absolute liability
		Sets liability limits for *willful* or *negligent* discharges
		1. vessels
		  –$10 million or $100/gross ton
		2. onshore installations
		  –$8 million
		3. offshore installations
		  –$8 million

What is not marginal is the position that oil district people hold on the House Committee. As of 1969 there were nine members who had been on the Committee since 1957. Four were from oil producing districts—Jim Wright of Texas, Ken Gray of Illinois, Frank Clark of Pennsylvania, and Ed Edmondson of Oklahoma. Of the nine, Wright and Edmondson are distinguished for both high levels of ability and acceptance by other members. Typical of the opinions offered on the abilities of Public Works members are the comments of a former BOB official. After going down the list of names of Committee members and citing the defects of each he came to Wright and Edmondson.

Wright, he's smart, aggressive, and tougher than hell. And Edmondson's the same way. They're quite effective together.

James Murphy, in his study of the Committee, also sees Edmondson and Wright as exceptional:

Within the House, only unusual ability—perhaps backed up by state delegation support (e.g., Wright, Edmondson)—can overcome the obscurity of the Public Works Committee member who has not attained a position of leadership.[73]

The efforts of Wright and Edmondson to defend the oil industry from unfavorable features of water pollution legislation are best exemplified in the 1969 bill. During the public hearings on H.R. 4148, Wright questioned several witnesses to establish his position and argue for more limited oil pollution legislation. For example, when Charles Teague, a House member representing spill-ridden Santa Barbara, testified on behalf of strong controls on offshore drilling, Wright commented:

Of course, there have been, I think, some 12,000 offshore wells in different points around the perimeter of the United States, and this is the only situation, I think, that has resulted in this kind of arrangement.[74]

Later he requested data on oil spills from the Interior Department. The data showed only thirteen major spills, most of which were clearly non-willful and non-negligent.[75]

When Secretary of the Interior Hickel testified before the Committee, Wright pressed him on Interior policy regarding offshore installations.

Wright: Now, is it a fair summation of your recommendation that with respect to those offshore facilities, which are there, after all at your sufferance, there will be absolute liability and without proof of fault that is with respect to vessels, the liability be related to the weight and size of the vessel and in no case more than $15 million as a requirement; that with respect to onshore facilities, which operate under the jurisdiction of the States and are not thereby your sufferance, you have no clear recommendation at this time? Is that a fair summation of the Department position?

Hickel: (Responds about control in lease and non-lease situation.)

Wright: Mr. Secretary, I am just trying to get a clean statement of the Department's position. Did I accurately state it?

Hickel: I think pretty good. On the offshore one where there is absolute liability, there does not seem to be any doubt, what we can do there and we are doing it. Whether we can do it with State jurisdiction onshore, whether we even want to, whether Congress would want to enact legislation that would say absolute unlimited liability in areas where they did not draw contracts with the people involved.[76]

To the casual observer this exchange may be trivial. But in the Committee's executive session, Wright referred directly to this exchange with Hickel to argue that liability on onshore leases should be limited. The idea for this kind of questioning came to Wright directly from oil industry officials. They were concerned over federal interference in areas of state control and voiced it through Wright.

He received support from Edmondson. Some of Oklahoma's oil production comes from land on Indian reservations. The leases for drilling are obtained from the government. Edmondson thought that an unlimited liability provision would make leases on Indian lands unprofitable in competitive circumstances. But the effect of Edmondson's and Wright's arguments was to protect the industry position on liability.

One may question whether these sentiments—especially Wright's, since he often sponsored water pollution legislation—demonstrate a constituency tie to the Congressman's issue position. An exchange in 1968 between Wright and Joe Moore, then Commissioner of the Federal Water Quality Control Administration, leaves little doubt.

Wright: . . . Anybody who has any kind of an installation on that river that uses oil of any kind, any industry, any type of operation that uses any oil you would come in and have controls over them, right?

Moore: Only if you had a spill on the waters.

Wright: I understand that. I understand that. I wonder why you seek this authority in respect to oil? I wonder why you are not seeking it in regard to these other pollutants your industrial waste profile series deals with? Why do you limit your request to oil? Petroleum products?[77]

In the next chapter we shall examine further why oil district representatives are especially successful with water pollution bills on the House Committee. The point to be made here, however, is that the constituency ties are effective at the committee level.

The Muskie Subcommittee is a different story. No oil state member has given serious attention to the actions of that Subcommittee. Sometimes Senator John Sherman Cooper of Kentucky would argue the oil industry position. Thus, in 1965, Cooper issued a minority report to S.4. He objected to the powers given the Secretary of Health, Education, and Welfare to set standards. But Cooper's

role was generally supportive of industry arguments against pollution legislation. True, Cooper represents Kentucky, which falls just outside the group of leading oil producing states, and some have claimed he makes no bones about protecting constituency interests in his state. The constituency tie, if it exists with Cooper, is not nearly as clear as with Wright, Edmondson, and other oil district representatives on House Public Works. One House committee member went down the list of current members, stopping as he came to each oil district Democrat to comment: "He went as far as he could. He's got constituency stakes to protect."

Compare this with an industry lobbyist saying: "When I've been at Senate hearings, sometimes there are only two Senators present. And one of them is Muskie. He's been in control ever since he got the Subcommittee."

We cannot leave this section without some comment on a subject to be explored more fully in the next chapter. Our main concern has been with water pollution legislation produced in the period since the Muskie Subcommittee was established. From our examination one can conclude that a reason for the relative success of the oil industry position in the House compared to the Senate has been the percentage of the relevant committee and subcommittee membership from oil producing districts and the effectiveness of their constituency ties. However, note that prior to 1964 the House produced water pollution legislation with stronger enforcement sections than the Senate. Yet, the Senate Public Works Subcommittee on Rivers and Harbors consistently had a lower percentage of members from oil states than the House Committee had from oil districts. Does this not cast serious doubt on the conclusion?

It would, except that the circumstances can be easily explained. During the 1955-1963 period, Robert Kerr of Oklahoma chaired the Rivers and Harbors Subcommittee. Given the low level of attention by most senators to subcommittee assignments on all but the major committees,[78] Kerr (like Muskie) controlled, to an enormous extent, the output of Rivers and Harbors. His attitude on water pollution legislation is described by Jennings as being one of "skepticism over an extensive federal commitment."[79] Throughout the 1955-1963 period "conservationists channeled most of their efforts through Blatnik and the House rather than through Kerr and the Senate."[80] Business and industrial concerns "consistently favored Senator Kerr's less expansive proposals if they had to make a choice between House and Senate versions."[81] After Kerr's death and the transfer of water pollution legislation to the new Muskie Subcommittee, where the percentage of oil state Senators was actually higher, the Senate produced stronger water pollution legislation than the House. Seemingly, this runs in the opposite direction of the conclusion. What it really symbolizes, though, is the need in the case of the Senate Committee to look at qualitative as well as quantitative levels of constituency support. The change of Subcommittee chairman from an oil state to a non-oil state Senator is more important than the number of oil state Senators on the Subcommittee.

Certainly, there are other differences between Kerr and Muskie. But of chief interest here is the fact that their constituency ties were opposite on the water pollution issue.

One might argue that the switch to producing the stronger bill in the Senate rather than the House may reflect something about the House Committee and not just change in Senate Subcommittee. But such an opinion has little foundation in fact. A major point made, in discussing the House Committee, was the stability of an important segment of its membership during the post 1955 period. Certainly there was a significant turnover in membership. The turnover was largely among freshmen Congressmen who served one term on Public Works and moved to another committee.[82] A substantial percentage of the Committee remained unchanged during the period. A longtime member of the House Committee, admitting that even with the high turnover rate there was little change in its make-up on the water pollution issue, went on to comment on the change from Kerr to Muskie on the Senate Subcommittee:

Muskie has a subcommittee that only has nine members. He can get anything through that he wants. . . . If he wants to do something he can. With Kerr it was the other way.

And later in the interview he claimed:

When Bob Kerr was on Public Works, the industry had a lot to say. . . . After Kerr died, Muskie took over the pollution area. . . . The whole thing changed.

In the next chapter this refinement will be elaborated. For now it is sufficient to note that, as in the tax area, the number of constituency ties the oil industry has to members of relevant committees is associated with variation in success of the industry's position on water pollution legislation. But other factors affect this simple relationship. Moreover, we have seen on the House Public Works Committee a clear case of overrepresentation of oil districts among its Democratic membership.

The Enforcement Agencies

Unlike the tax area, where there are clear ties between the oil industry and Treasury, such ties are much harder to document in the water pollution area. There are several reasons for this. First, the federal pollution control effort has moved from one agency to another with great frequency. Second, leadership in the enforcement efforts has regularly been composed of political appointees who have little substantive background in pollution control. Third, and most important, the impetus for and administration of enforcement and the creation of water quality standards, while under federal direction, is often left to states

and localities. Before examining the data on the industry's ties to people in the enforcement program, a discussion of these limitations is necessary.

The current federal agency in charge of water pollution enforcement is the Water Quality Office of the Environmental Protection Agency (EPA). It came into being in December 1970 hot on the trail of a long line of predecessors. In 1948 the water pollution control program was first embodied in the Division of Water Supply and Pollution Control (DWSPC) of the U.S. Public Health Service; in 1966 it became the Federal Water Pollution Control Administration (FWPCA) in the Department of Health, Education, and Welfare, which later transferred to the Department of the Interior; in April 1970, it became the Federal Water Quality Administration, also in Interior. With each change in name came a change in leadership. True, the staff of the program was fairly stable. Most, however, were technical employees—sanitary and civil engineers and some lawyers providing little direction for the program. One lawyer who has worked for nearly a decade for enforcement offices of the various water pollution agencies claimed that technical people such as himself simply act under the provisions and leadership they are given. Because of this he saw little interest in group activity.

Those pressure groups and lobbying groups would be felt at the political level of the agency. . . . Once we have locked horns legally there's no pressure here.

The political leadership, which is not particularly intertwined with industry interests, has been, with certain notable exceptions, either inexperienced, inept, or both. From 1961 to 1968 leadership of the water pollution control program was in the hands of James Quigley, a former Democratic Congressman from Pennsylvania who lost his House seat in the 1960 election. His only real experience with water had been as a Naval officer. His tenure is noted for his visits with friends on Capitol Hill, his vain preparations for a political comeback, and his general absence. The Nader report capsulizes Quigley's career with a story about one of his attempts to grasp the pollution problem:

When one of the technical people brought in a flow chart for the Commissioner to inspect, Quigley took it from the engineer, threw a little red pillow down on the rug, and lay down to look at it. As he turned the chart around, he asked, 'Where does it begin?' The engineer lay down on the floor with him to point out respectfully that Quigley was holding the chart upside down.[83]

After Quigley resigned in 1968 to become a vice president with U.S. Plywood, Joe Moore became Commissioner of FWQA. Moore has been the only person who came to this position with any serious experience with the water pollution control problem. In spite of the fact that he had previously headed the Texas Water Quality Board, which had substantial representation of industry groups, his behavior as Commissioner did not reveal any strong ties to industry

groups.[84] Perhaps this is due to the brevity of his tenure, a little more than one year, during which his major concern was creating order out of the administrative shambles left by Quigley.

In March 1969, as the Nixon Administration took over, Moore was replaced by David Dominick, whose experience with water pollution was limited to work he had done as a legislative assistant to Senator Clifford Hansen. According to the Nader report, Dominick admitted that he prepared Hansen's position of opposition to FWQA water quality standards. Dominick did, however, say that he favored such standards and was just doing work for Hansen.[85]

In any case, Dominick's experience was limited, and the greater part of his early service as Commissioner was spent learning his job and fighting with Carl Klein, Assistant Secretary (of the Interior) for Water Quality and Research.[86] With the movement of the water quality program to EPA, the bureaucratic mess finally appears to be cleared up.[87]

Even though there is a federal water pollution control program, the bulk of the enforcement effort comes from state and local officials. For example, the 1965 Water Quality Control Act gave the Secretary of HEW the authority to set water quality standards, but only if the states failed to do so; and then only indirectly by withholding federal funds. Throughout the history of federal water pollution legislation, a major argument against federal enforcement has come from industry groups—including the oil industry. While this argument carried decreasing weight over the years, the American Petroleum Institute still offered it as an objection as late as 1965. In a statement to the House Public Works Committee, API objected to Section 5 of S.4 because "a State's right to control its own water would be taken away, and effective cooperation and control would become impossible."[88]

There is a logical reason behind the industry groups' argument: enforcement from state and local officials was less likely than from federal officials, since state and local officials would be more concerned with the economic well-being of a local industry. But even beyond this logical argument, knowledge concerning the composition of state and local water pollution control boards affected the industry groups' position. Nearly all these boards had industry membership, if not control. In December of 1970 a *New York Times* survey of state pollution control boards discovered that the regulators and the regulatees were often the same people.

The roster of big corporations with employees on such boards read like an abbreviated blue book of American industry, particularly the most pollution-troubled segments of industry.[89]

Thus, industry groups naturally felt more secure working at the state level than with federal enforcement authorities. The industry's constituency ties and access to state and local officials are far stronger than those it has to federal

officials. In fact, the oil industry feels it has few friends at the federal enforcement levels and especially at EPA. When asked whether anyone at EPA represents the industry position on pollution issues, the response was vehement.

Represents? Certainly no one represents us. There are some people familiar with the industry and the way it operates but most of them think pollution is intolerable. Ken Biglane, for example, knows the industry very well and had a lot to do with the 1969 bill and setting the standards. But he believes that any spillage of oil should be corrected.

This is in marked contrast to the oil industry's access to the Treasury Department. But this does not mean that it ignores EPA. An EPA official has said that the industry is constantly trying to negotiate with EPA:

The industry works directly through EPA with Nick Gammelgard as their spokesman. They're especially concerned with Section 11 of the law which defines what a discharge of oil is. As yet, we have not been able to implement the regulation on this. But we'll meet with them and listen to their point of view.

However, the lack of a "reasonable" ear at EPA does not mean that the oil industry gets closed out once things reach the enforcement stage. This same EPA official claimed that the main problem EPA has had with setting regulations and enforcing provisions stems from the fact that the industry used its ties with other administration officials to appeal EPA decisions.

If they lose here then they go to the White House. Stans (Maurice Stans, Secretary of Commerce) has direct contact with the President, and sometimes we'll get the opinion of the White House on what we should do.

The need to sidestep the federal enforcement agency is relatively new. When the enforcement program was at Interior, the industry was somewhat more effective. One official who was with both Interior and EPA saw a big change.

Over at Interior we were just limping along. We'd institute some mercury suits or get after Chevron oil, but the actions were *ad hoc.*

However, Interior still holds much of the potential regulatory power—indirectly related to the water pollution problem—over the oil industry. Chief among these are decisions related to leasing offshore areas to oil companies for drilling purposes. Given the long-noted ties of mineral industries to the Interior Department, one would think that the industry would be comfortable in regard to drilling regulations. Surprisingly Walter Hickel, Nixon's first Secretary of the Interior, who was given a difficult time from conservation forces at the time of his confirmation and had ties to the oil industry, was considered tough by the industry because, as one official claimed, "he knew the industry and all their

tricks." It is not strange to find that Rogers Morton, who followed Hickel at Interior, was carefully questioned by mineral state Senators. According to this same official, "Allott (Senator from Colorado) was particularly skeptical." As a result, "he (Morton) goes out of his way to be favorable to oil." We must remember that Morton does not have a record of being anti-oil. If anything, the opposite is true. While on Ways and Means in 1969 Morton, whose Maryland district had some minor oil well drilling, stood firm with the industry on all committee votes.[90] But after the Hickel experience, industry officials were taking no chances. They could not, for instance, stand for delays in drilling and unlimited liability provisions on spills from wells on federal lands, as Hickel had proposed before the House Public Works Committee.[91]

Summary: Constituency and Water Pollution

All this just proves that constituency ties as broadly defined may have some impact, but they do not insure success on water pollution legislation. It helps to have Wright and seven other oil district Democrats on Public Works from 1965-1969 and Kerr as Chairman of the Senate Subcommittee until 1963, because they clearly assisted the industry position. But constituency ties in the pollution area did not insure, as it did in the depletion area, that decision makers would operate in response to the industry. Moreover, successes that the industry did have in this area were also due to positions taken by decision makers who were not constituency access points for the industry. Thus, Rep. William Cramer's (ranking minority member of the House Public Works Committee) anti-federal interference brand of conservatism was as important as Wright's crippling amendments. With the Muskie Subcommittee, where Republican Senators Boggs and Baker held a united front with Muskie, the industry had little impact. These relationships will be one focus of our attention in Chapter 3.

Despite the strong constituency ties of the industry at some decision points, its position did not always prevail. The 1965 or 1969 bills were not stopped. True, some serious modifications were made, but legislation with increasingly significant enforcement sections did pass. In later chapters we shall see why, in spite of these active constituency ties, the industry position gradually eroded.

Summary and Conclusions

Constituency ties are certainly an important factor in determining the success of an interest group in the policy making process. Our findings concerning the oil industry's ties in water pollution and depletion allowance legislation provide ample evidence that this is the case. In the tax area, it helps explain the consistently higher success rate of the industry with the Finance Committee—

and therefore the Senate in general—than with Ways and Means and the House. With water pollution legislation where, there is more variability in the strength of these ties, changes in their strength are associated with changes in the industry's focus of activity. There was little variability in the strength of constituency ties in the tax committees; therefore our claims of the importance of these ties could only be inferred from comments offered by respondents and the historical record on the subject. With the pollution area, not only do we have the comments of relevant actors but the switch of House and Senate Committee roles in the 1963-1964 period and the change in the Senate Subcommittee.

In addition, constituency ties to the executive agencies are important to industry success. This is demonstrated by the industry's overwhelming concern over leadership at the Treasury Department throughout the period and more recently its dissatisfaction with EPA's and Interior's (under Hickel) handling of the oil pollution matter. The industry urgently wants to have someone in an Executive decision making position who will insure "fair" consideration of the industry's position. There is consternation in the industry when this is not the case. The Surrey and Morton confirmation hearings are the most clearcut demonstration of this. The industry's use of Maurice Stans and state and local authorities to hinder the federal water pollution program, and its vocal displeasure with Hickel and Lyndon Johnson, who had been thought to be friends, are illustrative. In neither issue area does the industry view the legislative process as an activity limited to Congress. As we have documented, the industry sees the Executive agencies as important components in the process. Later we will discuss just how important the Executive agencies are in each policy area, but for now it is important to note the industry's concern with maintaining ties, its fears when the ties are not sufficiently strong, and its occasional lack of success when ties are weak.

The distress over Hickel and Johnson points out a necessary qualification we have had to make when discussing ties of the oil industry. Not all ties are effective. In determining the number of House and Senate ties to the industry, we were careful to include consideration of intensity and then to determine whether the representatives acted in accordance with these ties. With the Executive the same problem was faced. The finding that someone has ties with the industry either by representing or coming from an oil producing area, having business connections with the industry, or receiving campaign contributions from oil sources is not as valuable as being able to show that the decision maker acted in accordance with the industry's position. Exceptions like Paul Douglas, Walter Hickel, and Joe Moore were just that, however—exceptions, not the rule. Where some ties exist, the probability is that they were at least partially effective. (This finding is more true of the tax area than of water pollution, as will become clear in Chapter 4.)

A further refinement offered in this chapter is that the industry rarely had a majority of decision makers with ties to it at any one point in the legislative

process. This emphasizes the frequent need of the industry to build a winning coalition. In the depletion area, this took the form of expanding the coalition by logrolling other representatives from mineral areas. In the pollution area, a logroll, although used, was not as applicable. Instead the major effort was to stress state and local rights, interference of government in the affairs of business (economic liberalism), and most recently the energy crisis. Most of these efforts were designed to galvanize southern Democrats and Republicans with oil district representatives into the familiar conservative coalition. Although the conservative coalition cannot be seen in floor votes on pollution issues, it was certainly an active force in the House Public Works Committee, especially in the post-1964 era.

Both types of coalitions involved shifting the cleavage lines so that a minority oil interest could build a majority opposed to cutting the depletion allowance and to oil pollution legislation.

While we have stressed that constituency ties can be powerful, that at some important decision making points the industry possesses a disproportionate number of ties, and that within an issue area the strength of ties is related to industry success, we have withheld comment on the comparative success of the industry in the two issue areas. Most of the chapter shows that in each issue area the industry has many ties at some decision points and at others just a few. But the general impression one gets is that the industry has similar levels of constituency strength at decision points in the two areas. Thus, both the House Public Works Committee and the Senate Finance Committee have similarly high numbers of oil district members among the majority party, the Democrats. Although the industry appears to be in a weaker position at EPA than at Treasury, one must remember that state and local control of some pollution enforcement, as well as inefficiencies in the federal water pollution programs, compensate for this difference. If constituency ties were sufficient to explain variation in the success of the industry position, one would then expect equal success in the two issue areas. But this is hardly the case. While we will reserve full evaluation of the impact of legislation in these areas, it is clear that the oil industry has fared far better with depletion legislation than with water pollution legislation. Although oil pollution legislation was delayed, some of it has passed. Companies have had to insure themselves against spills and have assumed the responsibility for cleanup. Suits have been filed where companies have been negligent, even if under the 1899 Refuse Acts. Federal water quality standards are being established. Compare this to a marginal change in the oil depletion allowance from 27½ percent to 22 percent which, as we will show later, has had a much smaller impact than the percentage change would indicate.

One may claim that comparing success in these two issue areas is like comparing apples and oranges. But such is not the case; industry officials clearly see a difference between their success in stopping water pollution legislation and their success in preserving the depletion allowance. The same lobbyists, who talk

of the unreasonable demands of the Muskie Subcommittee and of EPA, take an entirely different tone when they speak of Ways and Means, Finance, and Treasury. They know there are people articulating the industry position at every decision making point in the tax process. The same view is held by the decision makers in each area. One Ways and Means member claimed, "if we met two more hours, Watts (Dem. Ky.) would have made it a tax relief bill instead of a tax reform bill." Another Democrat called the changes, especially those concerning the depletion allowance, "cosmetic not surgical." No such sentiments are echoed by the strong water pollution legislation advocates. They admit that they had a tough time passing the 1965 and 1969 bills, but with each defeat, they were able to up the ante. In the depletion area, opponents of the allowance kept returning with the same proposals for cuts—the Truman proposal of 1950 is remarkably similar to the Williams amendment of 1964—while each new pollution bill contained broader scope of coverage, higher liability limits, and increased federal authority.

There is a second difference in success that constituency ties at decision points do not explain—success over time. Why is depletion cut in 1969, when 41 percent of the Senate Finance Committee is composed of oil state Senators and 20 percent of the Ways and Means Committee by oil district Congressmen? Neither figure shows much variation with the 1950, 1951, or 1963 levels when depletion was maintained. The same case is harder to make with water pollution legislation, especially if we consider the pre-1963 period. Nevertheless, committee composition has been relatively stable regarding oil constituency representation since that time. Yet the oil industry's success is clearly on the wane.

When we compile all this evidence, we find that the level of constituency ties is in some ways related to the industry's success on each issue. At best, however, it operates in a haphazard fashion. It is certainly not a sufficient explanatory variable for understanding the degree of oil industry success. But just as clearly it is not an unrelated variable.

One final set of conclusions comes from this chapter. These relate to the Fenno-Froman claim about the impact of constituency size on interest aggregation in the House and Senate. Two basic refinements to this contention have been elaborated here. First, interests are not necessarily distributed in such a way that a greater proportion of Senators will have an interest in their constituencies than will House members. While it is correct that the average Senator will have more interests in his district than the average Congressman, interests are often distributed non-randomly. It should not be surprising to find, when studying any particular interest group, that a greater percentage of House members than Senate members represent districts containing this interest. Moreover, this question brings up a second complication—intensity. This point was previously elaborated, since it was vital to operational definition of an oil district.

We now turn our attention to other factors in the legislative process, rules and

procedures, which will assist us in qualifying the findings about the relationship between constituency strength and industry success. In addition, we shall check on other independent impacts that rules and procedures have on oil's success in our two issue areas.

**Rules, Procedures and
Processes**

Rules, procedures, and processes are rarely, if ever, neutral.[1] In the congressional process, they work to the advantage of some over others. We have seen the Senate cloture rule delay or defeat civil rights legislation, Supreme Court appointments, and the supersonic transport (SST).[2] The exigencies of committee chairmen worked to affect legislation on federal aid to education, medicare, and school prayer.[3] Committees have been stacked in attempts to free legislation.[4] Floor leaders have manipulated scheduling to strengthen their side on issues.[5]

Major fights on substantive legislation have often involved fights over changes, rules and procedures. A classic example is provided in the case of the "21 day rule." Liberals realized that much of the legislation they favored bogged down in the Rules Committee. During the 81st Congress, they had used a "21 day rule" to bring bills to the House floor from an unresponsive Rules Committee. After the 1950 election liberal ranks in the House were depleted, and the rule was dropped. It was not until the start of the 89th Congress that liberals had sufficient strength to institute a new "21 day rule." During the 1950-1965 period many programs, having sufficient support on the House floor, failed to reach it for lack of a ruling from the Rules Committee. During the 81st and 89th Congresses many major pieces of legislation were brought to the House floor by using a "21 day rule."[6] But as in 1950, the 1966 election resulted in losses in liberal House membership, and the "21 day rule" again was deleted.

Clearly the existence or non-existence of a "21 day rule" had significant impact on the success of liberal policies in the House. It may be argued that all that changes in a "21 day rule" reflect is the numbers of liberals in the House.[7] While this is true, it cannot be used to deny the independent impact of rules on policy. In 1965 liberals had the strength in the House to make the rule change but did not have the strength to bring certain pieces of legislation to the floor without the rule change. Thus, rules and procedures may reflect more than just the strengths of competing groups of Congressmen. Despite their numbers in the 89th Congress, liberals needed a "21 day rule" to counteract other existing procedures.

Rules and procedures, in affecting the course of public policy and the actions of Congressmen, have a simultaneous effect on interest groups concerned with various pieces of legislation. The way the process operates has a good deal to do with the success of an interest group's position and where interest groups will center their attention in attempting to influence those results.

Most importantly, the advantages that rules, procedures, and processes offer are not randomly distributed, but in fact work systematically to favor groups holding certain types of positions on given issues. One basic contention to be set forth in this chapter is that rules, procedures, and processes have a built-in non-decision bias, and further that this bias affects the success the oil industry has had in each of the two issue areas under study.

"Non-decision" is the term used by Bachrach and Baratz:

A decision that results in suppression or the wanting of a latent or manifest challenge to the values or interests of the decision-maker. To be more explicit, non-decision-making is a means by which demands for change in the existing allocation of benefits and privileges . . . can be suffocated before they are ever voiced; or kept covert, or killed before they gain access to the relevant decision-making arena; or, failing all these things, maimed or destroyed in the decision implementing stage of the policy process.[8]

However, the second purpose of this chapter is to show that, despite the common favorable impact of rules, procedures, and processes on the industry's position in each issue area, their operation and impact are somewhat different on the tax issue than on the water pollution issue. Or to phrase it more simply, the tax legislative process and the water pollution legislative process, while having certain common features, each possess some degree of institutional uniqueness. Moreover, these differences affect the activities of the industry and are to some degree responsible for variations in the industry's level of success in the two issue areas.

It is impossible to examine all the facets of the Congressional process bearing on each of these issues. Our attention will be on the most prominent and most important of these. Much of our focus, therefore, as in the preceding chapter, will be on the relevant Congressional committees. Specifically, we will examine the recruitment of members to these committees and how the committees vary in their operations. The former will consider why some committees have a disproportionate number of oil district members, whether the industry has tried to influence recruitment, and the impact that changes in recruitment procedures have had on policy outputs. The latter's concern is with how decisions are made by these committees—who has control, who participates, and how it affects the places where the industry decides to make its case.

The examination of recruitment to the committees and committee operations will be placed in the broader context of the common features that work to the oil industry's advantage on both issues. Thus, the chapter begins with a study of the assistance the oil industry receives by being on the defensive side on each of these issues. Then, after a study of committee recruitment and operation, the impact of the incremental nature of Congressional policy making will be analyzed. These two broader features, together with the variable ones sandwiched in between, will provide the context for the contention that the net

impact of these features is to favor non-decision politics, which is precisely the oil industry's goal in each of these issue areas. Moreover, throughout the chapter it will be demonstrated how the oil industry has used these advantages to prevent legislative action on the two issues.

A final section will then draw this together with some other features of the Congressional processes to summarize the case for the built-in non-decision bias.

Advantage to the Defense

In Chapter 1 it was stated that the selection of issue areas for examination was based in part on a desire to exercise certain controls. One of the controls exercised in selecting the two issues was the industry's defensive position in the legislative process. That is, on both depletion allowance and water pollution legislation the industry opposed most major legislative proposals. There is a vital reason for this control. The defending side in the legislative process has a major advantage over the initiating side. Prime among these is what Froman describes as the "serial" nature of the legislative process.[9] In brief, his proposition is that the proponent side in the legislative struggle needs to build a winning coalition at each of the many decision making points in the process. The opponents, however, often need only to win at one decision making point to delay, if not defeat, a legislative proposal.

What this means for interest group activity is that a defending group can, if it desires, concentrate its efforts and resources on a single decision making point while the proponents may be spread thin. (Of course, since the legislative process is serial in nature, this resource disadvantage may be exaggerated.)[10] More importantly, defending groups that lose at one point can come back to fight at the next decision point. For them, each decision point is only a battle. For the proponents, losing a battle often means losing the war.

Therefore, the first point to note about the impact of rules and procedures on the oil industry's level of success in these two issue areas is that being on the defensive gave it definite advantages over the proponents of legislative change. And when the industry was placed in a proponent position, the success of its position markedly decreased.

The Depletion Allowance

The depletion allowance fight illustrates the above propositions in a most interesting fashion. In 1950 and again in 1951, when the Truman administration proposed the cut in depletion to 15 percent for oil and gas, the House Ways and Means Committee was the next decision point. As we discovered in Chapter 2, the level of industry constituency ties with Ways and Means members was lower

than with other decision making points, especially among majority Democrats. In 1950 and 1951, the Committee faced a barrage of pro-depletion testimony from industry spokesmen and Congressmen from oil districts. Accompanied by the major logrolling campaign cited earlier, the strategy proved successful. Instead of lowering depletion for oil and gas, the Committee reaffirmed its position and added new substances to the group receiving percentage depletion.

Having been successful in Ways and Means, oil industry lobbyists paid only lip service to the rest of the process of H.R. 4473, the 1951 tax bill. They knew that the closed rule, which prevented amendments on the House floor, would protect depletion there and that the Finance Committee of the Senate was more favorable to depletion than Ways and Means. When H.R. 4473 reached the Finance Committee, the oil industry set its attention to the excess profits provision of the bill and totally neglected to comment on depletion.[11] Correspondingly, Secretary of the Treasury John W. Snyder, the Truman administration spokesman, testified before Finance without mentioning depletion allowances.[12] Once the industry position was upheld by Ways and Means, it was safe and the administration proposal unsalvageable.

The 1963 tax bill presents a clearer example of this phenomenon. President Kennedy's proposal for a tax cut included indirect methods of cutting the tax advantage that the depletion allowance gave the oil industry. The impact of his proposals would have had the same result as a reduction in the depletion allowance to 22 percent. Ways and Means, after considerable industry activity, failed to accept three of four administration provisions on the treatment of mineral resources. On October 16 of that year Senator Williams of Delaware, during a Finance Committee hearing, chided Secreatry of the Treasury Dillon on the administration's position:

Senator Williams: Is the administration fully satisfied with the changes which this bill makes in connection with the oil depletion or do you recommend that we do something further?

Secretary Dillon: No, we are not fully satisfied at all. We made a further recommendation, but after a long discussion this was turned down. It was turned down by a very substantial vote of the House Ways and Means Committee, 3 or 4 to 1, and so I see no possibility of our recommendations being accepted and going in the bill, so we are not making suggestions that that be reopened . . .

Senator Williams: Some of us are naturally born optimists. Would you submit to this committee a draft of your recommendations so the committee could consider them in our deliberations? I mean a draft of the language.

Secretary Dillon: We never made any language draft because, as you know, the Constitution provides that tax bills originate in the House of Representa-

tives . . . , and the language is all drafted in the House itself by its own drafters. . . .

Senator Williams: The Constitution also provides that the Senate consider amendments?

Secretary Dillon: Yes.

Senator Williams: And upon our request you can furnish suggested language for such an amendment?

Secretary Dillon: That is correct.

Senator Williams: So will you furnish that?

Secretary Dillon: If you request that, we would be glad to furnish language to you for whatever use you want to make of it that would carry out the recommendations we originally made. But we want to make it clear that we think that it would be a waste of time to get into detailed consideration of this particular problem at this time.[13]

Dillon was simply claiming that the battle and war had been lost in Ways and Means and that nothing was to be gained by putting a dead horse back on its feet.

The 1969 tax bill provides an interesting contrast to these cases. For the first time the industry had to go from a position of opposition to one of advocacy. After Ways and Means voted to cut the depletion allowance on oil and gas to 20 percent, the industry expected to recoup its losses in the Senate Finance Committee. The bill Finance worked with was the House-passed bill, which included this cut. Therefore, for the first time, the industry was placed in the quasi-proponent position of asking for restoration to 27½ percent. True, deleting the reduction provision at one decision point might be sufficient to avoid any change. (Of course, the industry now had to concern itself also with the action of the conference committee.) But the industry position took on a new burden that was not fully recognized until the Finance Committee voted. The industry had expected to win restoration to 27½ percent, by a 10 to 7 vote in the Finance Committee. But Eugene McCarthy (Dem., Minn.) was "unavailable" and Vance Hartke (Dem., Ind.) announced the day before the vote that he favored a cut to 15 percent.[14] The vote taken on October 23 resulted in an 8 to 8 tie.[15] Since the move to restore depletion to 27½ percent was an amendment to H.R. 13270, the measure failed. Had the effort been one by tax reformers to cut depletion to the House-passed level, the industry position would have been safe. Howeever, the Finance Committee was working with the House bill and not the existing tax law.

Still, the industry was only in a quasi-proponent stage. The Finance Committee voted to restore depletion to 23 percent and a further appeal could

be made on the Senate floor. In this sense the industry maintained a defending position in trying to keep the depletion provision as it existed in the tax code. Faced with the problem of restoring the old rate, the industry lost some of the advantages the process provides for a pure defending position.

It is not the claim here that proponent groups cannot also use the "serial" process to their advantage. They can have provisions inserted into a bill that were left out at previous decision making points.[16] Their problem is that defeat or delay of the bill means defeat or delay for the provisions they desire. For the oil industry, defeat or delay of H.R. 13270 would have been a victory.

Water Pollution

The conflict over water pollution legislation also demonstrates successful use of the serial decision making process for delaying legislation. This is especially true of the 1966-1970 period, when the industry held off a major part of its effort until very late in the process. In early 1967 the Justice Department announced that the 1966 water pollution bill had crippled prosecution of oil pollution from vessels. The Department assigned the blame for this to an amendment added in the conference committee session by Jim Wright of Texas. The amendment added only one word to the bill. Instead of defining a discharge as "any negligent or willful spilling" of oil, the amendment defined a discharge as "any grossly negligent or willful spilling" of oil.[17] The Justice Department was forced to prove gross negligence while relying on the ship's crew as the only witnesses. When questioned as to his reasons for including the amendment, Wright said that it was to satisfy industry objections and to protect the industry.[18]

In addition, the oil industry reportedly engaged in active lobbying during the 1966 conference to insure that the Senate bill provision covering shore installations did not become part of the final bill. The industry used the conference to negate the two provisions of the bill it found undesirable. Either the original Senate- or House-passed bill would have been stronger than the bill reported from conference in dealing with oil pollution.

It took until 1970 to correct the inadequacies of the 1966 bill. Each year there were new bills attempting to deal with the oil pollution problem. In 1967 Muskie offered S. 2740. It made any oil pollution open to civil penalties. But the House did not hold hearings. In 1968 both the House and Senate passed versions of S. 3206 twice. Extensive lobbying by the oil industry is credited with preventing the House from accepting Senate provisions on oil pollution, and adjournment arrived before a compromise could be reached.[19] H.R. 4148, the Water Quality Improvement Act, like the three bills that preceded it, had major sections dealing with oil pollution. Supposedly oil interests had given up on trying to stop the bill completely and were merely concerned with the extent of liability.[20] The House and Senate bills differed liability limitations. The bill

went to conference on October 9. Not until March 23 did the conference come to an agreement, and then only after oil spills occurred off the Florida and Louisiana coasts. One spill from a Humble oil tanker took place off St. Petersburg, Florida just prior to agreement. Cleanup was delayed because the federal government lacked the authority to charge the cleanup costs to Humble. St. Petersburg just happened to be represented in Congress by William Cramer, the ranking minority member of the House Public Works Committee and a candidate for the Senate. Cramer, in both public statements and executive sessions, was opposed to unlimited liability or liability beyond insurable levels in the case of oil spills. Further, he was an active, aggressive participant who delayed Committee action on water pollution by lengthy questioning of witnesses and by introducing many technical amendments during executive session. Democrats on Public Works claimed that Cramer even went so far as to keep all Republicans but one from attending executive sessions of the Committee. With some absences on the Democratic side, the one Republican would prevent the Committee from acting by citing the lack of a quorum. After the St. Petersburg spill, Cramer is supposed to have moved from his adamant position, allowing the conference to reach a compromise closer to the Senate bill. A staff member on the Senate Subcommittee claims: "It turned Cramer around." It is questionable, given the amount of public pressure for the bill prior to the conference, whether the conferees could have reached agreement if it were not for Cramer's delicate position.

These examples are not intended to show that the industry does not become active until the conference stage. This is anything but the case. As discussed in Chapter 2, the industry was active at earlier stages. Here, the purpose is to illustrate that (1) the defending position of the industry allowed it to delay or defeat pollution legislation even at late points in the decision making process; and (2) the concentration of its effort may come when the other side appears to have won. Moreover, a valuable contrast is available from this examination of industry lobbying in the two issue areas. On tax legislation, with the exception of 1969, the industry position was successfully maintained in the early stages of the process, and that was often sufficient. As one oil lobbyist put it: "We ease off to an extent, but we don't take any chances. People in the Senate friendly to us influence what we do. If they say 'We don't want a big show,' we listen to them." On water pollution legislation the industry's position, i.e., preventing legislation that would place responsibility for spills on the oil industry and its affiliates, did not succeed until the end of the process. Small victories were won when the House Committee reported considerably weaker bills than the Senate Committee. But it was not until bills were passed in both bodies that the industry position was saved. In Chapter 5, we will examine this phenomenon more fully. However, it should be noted here that this is indicative of the different operating styles to which lobbyists must adjust when dealing with an issue area that is highly material as opposed to one that is highly symbolic. In

both cases private interest groups will be most effective at low visibility stages in the legislative process. But to the degree that a given issue fits the pattern of symbolic politics, success for the position of small, organized, private groups is likely to come late in the Congressional process if at all. To the degree that the issue has overriding material consequences for the small, organized private group, success of the group's position is likely to be early before the conflict becomes fully socialized.

Committee Recruitment

In the previous chapter, it was demonstrated that on depletion allowance legislation, the oil industry position fared better in the Senate than in the House, while on water pollution legislation the reverse was true. This result corresponded to some degree to the data on the number of oil district representatives on the relevant committees handling the legislation. Naturally, we are directed to the next logical question. What types of people get on these committees? How are they chosen? And do the recruitment procedures of the committees affect the success of the oil industry's position?

Ways and Means

Since Republicans on the committees tend to take a consistent pro-industry position on both issues, our main concern is with recruitment of the Democrats.[21] Ways and Means recruitment presents the most interesting case. While Sam Rayburn was Speaker of the House, it has been alleged, no Democrat needed apply for a seat on Ways and Means unless he were safe on the oil depletion allowance. John Manley in his comprehensive analysis of the Committee notes:

It is generally believed that Rayburn would not let anyone on Ways and Means who would vote to cut the 27½ percent allowance given to the oil industry, a provision that has become the symbol of tax loopholes and the *cause celebre* of liberal tax reformers, but not one member who got on the Committee under Rayburn could recall being asked if he supported the oil depletion allowance. But if Rayburn did not normally interview the members on the oil question he did allow the perpetuation of the widespread belief that no depletion reformers need apply, and this was sufficient. One of his closest associates said unequivocally: 'Rayburn had two things. One was trade. He was 100 percent for international trade. And the second was the depletion allowance for oil and gas. There were the two.' And when Richard Bolling wanted a way to turn down Rayburn's request that he go on Ways and Means he found that all he had to do was point out that he would vote to reduce the depletion allowance.[22]

Manley properly hedges on the depletion case. He later intimates that it was more important that a member of Ways and Means not be a rebel or crusader against the depletion allowance.[23]

My interviews confirm Manley's findings. One member said he was not asked about his position on oil, but Rayburn was aware of his sympathy with mineral industries from work he had done on the Mineral Subcommittee on the Interior and Insular Affairs Committee. He added that he believed that Rayburn would not exclude someone for his position on oil but looked for men with "judgment." Another member claimed that his main difficulty in getting on the Committee was the trade issue and not depletion, on which he was known to be safe.

As much as people discount the importance of one's position on the oil depletion allowance to getting on Ways and Means, other evidence indicates that while Rayburn was Speaker the issue served as a major selection criterion. It is difficult to explain in other terms the seeming lack of anti-depletion statements or questions from Committee members during Ways and Means tax hearings prior to Rayburn's death. This holds for oil and non-oil district members alike.[24]

But Rayburn died in 1961, and this appears to have had a significant effect on the Committee's position on the depletion allowance. When asked whether Rayburn's death had an impact on the eventual change on depletion in 1969, Committee members offered conflicting opinions. A sampling of comments gives some idea of the lack of agreement on this subject. Three Republicans offered varied points of view:

A Senior Republican: While Rayburn's influence on appointments had been significant, it is only a marginal factor in explaining why the allowance was cut.

A Junior Republican: The change in the Committee's membership made little difference. No one goes on Ways and Means with the express purpose of cutting the depletion allowance.[25]

A Senior Republican: Rayburn's death meant a loss of control of assignments. That's the biggest reason for the change.

Democrats were no more in agreement than Republicans in responding to the question. One southern Democrat reflected varied points of view within his own view. He was quick to point out that anti-depletion members such as Charles Vanik were unopposed in their efforts to get on Ways and Means and that Rayburn, had he been alive, would have had difficulty stopping him.[26] But shortly thereafter he noted: "There has been a growing minority interest on the Committee that favored cutting the allowance. . . . They (Corman, Gibbons, and Vanik) probably would prefer to do away with it entirely."

The lobbying community was no more in agreement on Rayburn's impact than were the members of the Committee. One oil lobbyist flatly claimed, "The thing about Rayburn and depletion is a myth," while another felt Rayburn had at least slowed the change down a little.

Our data, however, provide support for the notion of a growing minority interest in cutting depletion among Committee members appointed after Rayburn's death. Table 2-5 in Chapter 2 shows that seven members voted for the Gibbons motion to cut the depletion allowance for all minerals by 40 percent—thus lowering oil from 27½ percent to 16½ percent. (As will be demonstrated in Chapter 5, the Gibbons motion was the only one that would make a major difference in the tax situation of most oil producers.) Of those seven, all Democrats, six joined the Committee after Rayburn's death and the seventh, Congressman James Burke of Massachusetts, became a member in 1961 only a few months before Rayburn's death. True, only four of the eight other Committee Democrats had been members before 1961. Nevertheless, it is significant that a committee that in 1960 had not a member who vocally opposed the depletion allowance should by 1969 have seven members willing to vote for a major cut in the allowance.

Additional support for this change in recruitment to Ways and Means comes from two of the Democrats who voted for the Gibbons motion. One explained that his state delegation was supporting another member of the delegation for the vacancy, but that he went outside the delegation for support:

I wrote letters to all the House members. To which thirty or so responded with support immediately and to which Boggs replied 'I'm for you. . . .'
At no time did anybody ask me about the depletion allowance or anything else. When I talked to Albert and McCormack, they never mentioned it. . . .

The respondent admitted that he had made speeches against the depletion allowance before he was appointed to Ways and Means:

Everybody clearly got the message that I could not be counted on as an oil vote. I had some trouble with the Texas delegation. But I think I had full support from Louisiana. . . . I guess the oil people thought they had so many votes they weren't worried about me.

The other member's story is somewhat different. Again, no one asked him about his policy position. But when a vacancy occurred, he went to the more senior members in his state delegation to inquire if any of them desired the Ways and Means seat. None of them did. Further, no one in the delegation had the same level of seniority as the respondent, and, he claimed, no one with less seniority would challenge. Since his state always has a Ways and Means seat, it was a strict case of seniority. He did qualify his answer, however, by suggesting that if someone with equal seniority in the delegation was favorable to oil and wanted to get on Ways and Means, that person might have been given preference over him.

Although one cannot conclude that the results in these two cases would have been different if Rayburn were around, the data suggest that holding the correct position on the depletion issue was no longer a criterion for appointment to Ways and Means after Rayburn's death.

Manley provides additional evidence for this change. He cites the case of a man who felt that Rayburn prevented him from getting on Ways and Means because of his position on oil. This individual finally got on the Committee when, he claimed, he had "toned down" his position.[27] More important for our examination is that this man, who was denied the seat when Rayburn was alive, got it after Rayburn's death.[28] While it is difficult to evaluate the importance of each of these factors to the man's appointment, it is worth noting that the mention of his name often brought negative comments from the Committee members I interviewed. Typical is the comment by a Committee Republican in discussing the 1969 tax bill: "He just wanted the publicity."

The alteration in the Committee caused by this recruitment change can clearly be seen in the turnover in two seats just prior to the 1969 tax bill. The first involved a switch in the California seat from Cecil King to Jim Corman, and the second in the Florida—the southeast district seat, from Sid Herlong to Sam Gibbons. A former Treasury Department official claims that these switches were major factors in producing the change in depletion. Commenting about King, he said, "He would vote with the administration on everything but oil. He felt he had to protect the independents—like Rayburn did." Herlong, who obtained a seat on Ways and Means despite the lack of Speaker Rayburn's support, was nevertheless a devout supporter of depletion allowances.[29] Gibbons, on the other hand, was a vocal opponent of them even prior to his filling the vacancy left by Herlong.

The change in the Committee's membership was not sufficient to cause the 1969 change in the depletion allowance. It was, however, a necessary factor. Without the change, Ways and Means might have remained the dead-end for any proposal to lower the allowance. With the exception of amendments on the Senate floor prior to 1969, depletion became a non-issue once Ways and Means hearings were held.[30] By 1969, a significant vocal minority had developed on the Committee that would not allow a tax reform bill to go to the floor without including a change in the depletion allowance. If it did, they would make sure it would lose credibility as a reform bill. They did not have the votes to carry the Gibbons motion, nor did Boggs have to go as far as 20 percent in order not to lose, but they did have the ability to raise a fuss and keep depletion alive as an issue.[31]

Finance

Ways and Means is not the only committee where recruitment practices affect the success of an interest group's position. It happens to be a valuable example,

because recruitment practices changed and with them changed the fortunes of the oil industry's position during the period under consideration. Moreover, this change was independent of a change in the number of potential constituency ties the oil industry had with members of the Committee. But even without an alteration in recruitment, we can see other examples of recruitment practices having favorable results for the oil industry. In the case of the Senate Finance Committee the effect is clear. Since Truman's attacks on the depletion allowance, both parties have consistently overrepresented oil states on the Finance Committee (see Table 2-15 in Chapter 2). For the Democrats, at least, this does not appear to be an accident. One Senator commented that the appointment process for Finance was more favorable to extractive industries than was the one for Ways and Means:

Johnson stacked the Committee. Probably Rayburn did the same thing in the House but not as easily. . . . That committee (Finance) has been loaded on oil for years.

According to this man, Lyndon Johnson had committee assignments made before he went to Steering Committee meetings. Support for this position is available in the Rowland Evans and Robert Novak book on Johnson. They relate a story of a phone conversation between Johnson and Mrs. John Stennis in which Johnson congratulates her on her husband's appointment to the Senate Appropriations Committee. He tells her, "The Senate elected your husband to the Appropriations Committee," and goes on to tell her how the Steering Committee selected Stennis for the assignment. But Evans and Novak are quick to observe "implicitly he was belaboring the obvious—that it wasn't the Steering Committee or the full Senate that really was responsible. It was L.B.J."[32]

A Senate staff member has been quoted as saying that Johnson went on the Finance Committee in 1955 to block Paul Douglas' request for the committee. Depletion and tax reforms had been Douglas' whipping boy.[33] Douglas confirms this in his memoirs. He claims:

For several years I wanted to be transferred from the Labor to the Finance Committee, where future tax policy was usually determined. But Johnson and the Steering Committee always passed me by. First, Russell Long and Clinton Anderson were put on the committee; then, Barkley. Upon Barkley's death, Johnson assigned himself. Smathers, although two years my junior, was also advanced over me. When, in 1955, another vacancy occurred and I was senior to all other applicants, I again applied. I got nowhere. The liberal commentator Doris Fleeson then began a series of articles in which she accused Johnson of opposing me at the behest of the Texas oil and gas interests. This was probably true. For months he resisted Miss Fleeson, but he finally gave in and allowed me to take a vacant seat. After all, I would be only one of fifteen members.[34]

After Johnson became Vice President, his role in influencing Senate procedural decisions weakened considerably. But the industry itself picked up the

slack in influencing Finance appointments. In 1961 William Fulbright was persuaded to give up his seat on Banking and Currency "by a combination of oil and gas interests in his state who wanted his voice in Finance and the Senate grandees who wanted to block Proxmire from getting on the committee."[35] The reference to "Senate grandees" no doubt refers to the Democratic Steering Committee, which formally makes committee assignments. This Steering Committee has traditionally been more conservative than the Democratic Senate membership.[36] Even after Johnson left the Senate and Majority Leader Mike Mansfield shared with the Steering Committee the responsibility for making committee assignments, Finance continued to overrepresent conservative, oil state interests.[37] As late as 1971 the attitude of the Steering Committee was reflected in a vote filling a Finance vacancy. Lloyd Bentsen, Jr. of Texas, a freshman Senator, lost out to Gaylord Nelson, a veteran of eight years in the Senate by only one vote. National Journal quoted one lobbyist as saying "the oil and medical people both were looking for friends."[38]

Similarly, the Republican committee assignment process in the Senate has been open to interest group influence. Stephen Horn finds testimony to the selective nature of this activity:

One longtime Republican participant in the Committee on Committees discussions stated although he had never known of interest group activity on an appropriations appointment, 'they become extremely active when selections for Commerce or Finance are under consideration.' Where one vote might tip the scales for or against revising the oil depletion allowance or . . . , an assignment becomes a 'life or death' matter, and lobbyists actively seek assignments for 'friendly' Senators.[39]

When questioned, industry officials denied taking an active role in the committee assignment process. They claimed that such activity would interfere with the internal workings of the House and Senate and would make them more enemies than friends. They made no effort to hide the fact that friends of the industry in Congress knew which candidates the industry preferred, and that it was unnecessary for them to make a public show.

It makes little difference whether the industry is formally active in this area. Until 1961 the recruitment process to both Ways and Means and Finance Committees worked informally to favor people who at least were not hostile to the oil industry. After 1961, recruitment practices for Ways and Means changed. House Speaker John McCormack of Massachusetts took a less active role in the filling of vacancies than Rayburn had. And when the Democratic leadership did become active, it was for the express purpose of stacking the Committee with members favorable to Medicare.[40] Clearly for McCormack, depletion was not the sacred cow that it was for Rayburn.

Even though the Senate Finance Committee's recruitment process may have become less autocratic when Mansfield became majority leader, few vacancies

occurred during the 1961-1969 period, and replacements continued to show an oil state bias. Six of the ten Democratic members in 1969 were on the Committee prior to 1961 and of the replacements, only one, Ribicoff, was clearly opposed to depletion.[41] Compare this to Ways and Means Democrats where only five of fifteen Democrats on the Committee in 1969 had been appointed prior to Rayburn's death in 1961. Thus, the change in Ways and Means' position on the depletion allowance was not just a function of a change in the recruitment process but required, in addition, that a substantial number of vacancies occur. If there had been a change in Finance recruitment, a doubtful proposition, the turnover rate was sufficiently low that little impact would have been felt.

The Public Works Committees

When recruitment practices of the House and Senate Public Works Committees are examined, one is struck by the way they contrast with the practices of the tax committees. Perhaps the term "benign neglect" best describes the recruitment to the Public Works Committees. There is no question of candidates being screened on their policy positions before going on these committees. They are not exclusive committees and do not attract the degree of attention from the leadership or from interest groups that Ways and Means and Finance do when time comes to fill vacancies.

James Murphy has noted this point in discussing membership of the House Committee:

Candidates seeking assignment to Public Works are not screened for party regularity or policy views. Though there are isolated instances of denial, the party leadership does not generally use them to promote policy or reward party regularity. Congressmen who wish to be assigned to Public Works need only the availability of a vacancy and sufficient seniority within their state party delegation to bump other possible candidates.[42]

While this criterion would seemingly preclude bias—either favorable or unfavorable to the oil industry—it does not happen to work that way. Certain states regularly keep members on the House Committee to serve their interests in public works projects. Between 1955 and 1969, Texas always had two members on the Committee and Oklahoma and Louisiana one each. This is not because these states had a particular interest in protecting the oil industry against pollution laws, but because of the importance of the day-to-day work of the Committee in the reporting of bills for road projects, harbor dredging, dam construction, and public building construction. Public Works is a constituency-oriented committee. Pollution legislation is not a main source of its attractiveness or attention. The fact that the Committee overrepresents oil producing

districts is not due to any effort by the leadership to stack it. It is rather the result of a coincident interest of oil producing states in public works projects. Thus, leaving aside specific policy considerations that may affect the recruitment process, we have the basic residual factor: who requests certain assignments has a lot to do with who obtains the assignment.[43] House members may request certain committee assignments for one set of policy reasons and affect policy unrelated to those reasons. It is like the comedian's opening line, "A funny thing happened to me on my way to the studio. . . ." A House member may request a certain committee assignment to serve his constituency in a particular fashion and discover that other things, which had nothing to do with his choice, are available to him once he arrives. Such is the case with membership on the House Public Works Committee.

This, of course, is not the only effect of "self-selective recruitment" on the oil industry position or water pollution. A factor cited earlier is that Congressmen tend to transfer off Public Works at a high rate. However, some Democrats make their career on Public Works. During the 1955-1969 period, twelve Democrats served at least ten consecutive years on the Committee. Five represented oil producing districts.[44] The other seats showed high turnover levels among Democrats. For example, during 1959-1965, there were twenty-one new Democratic members of this Committee. Only five of them came from oil districts. What this reports is a tendency among Congressmen who come from oil districts to stay on the Committee and not to transfer. Again, it may just be that members from states that always have membership on the Committee tend not to transfer at as high a rate as those from other states. The result, however, is to produce a Committee on which oil district representatives are not only overrepresentative of their number in the House but also overrepresentative of their number in senior positions on the Committee.[45]

Much of what is true about recruitment to the House Committee is also true of the Senate Committee. There is no active effort to deny anyone a seat or to press members into service. But among Democrats, most non-freshmen also hold seats on more important committees. States do not try to keep a member on the Senate Committee, as they do on the House Committee. At least there is no evidence of this.

The pattern here differs significantly from that regarding the Finance Committee. Not only is there a definite interest in who gets on Finance, but some members of the Committee are fairly active. They find the work of the Committee an important order of their business in the Senate. True, Russell Long is often credited with having substantial impact on what Finance reports, and his influence stems in part from the fact that the Committee is not a focal point of attention for some members. David Price found, in his study of the Finance Committee:

Many Finance Committee members were much more active legislatively on their second committee than they were on Finance. Douglas, Metcalf, Ribicoff,

Talmadge, Smathers, Fulbright, Dirksen, and even Anderson were cases in point. A partial explanation in the latter four cases was the duties which being chairman or ranking minority member of another committee entailed.[46]

However, Gore, Harris, Talmadge, Williams, Bennett, Hansen, and Miller were active participants in the Committee's deliberations on the Tax Reform Act. Moreover, Finance, like Ways and Means, had no subcommittees to institutionalize divisions of labor and specialization.[47] Public Works is not all like that. For most members it is of secondary importance. If they pay real attention to Public Works, they do so largely to insure projects for their states. For this purpose, the Subcommittee on Air and Water Pollution is least suited. Often Committee members will concentrate on one subcommittee for a few years and then go on to another. A staff member to one Public Works Senator commented that the Senator had left the Roads Subcommittee recently and gone on another "because ____ (his state) had enough roads projects by now."

Thus, our two issue areas show two different patterns of recruitment. In the tax area the industry has taken an active interest in the choice of new members for the two committees. Even on Ways and Means, where oil districts were underrepresented throughout the period under study, recruitment was used to maintain support for the depletion allowance. Only after Rayburn's death did a change occur, and the effect of that change in recruitment was finally felt in 1969. By then a vocal minority of depletion allowance opponents were able to make depletion an issue and, in 1969, force the Committee to report out a bill containing a lowering of allowances. No similar change has yet taken place with regard to the Finance Committee. But there, as the Bentsen case illustrates, the industry is still actively involved in recruitment of new members.

The two Public Works Committees show what happens when an interest takes little concern with recruitment. The House Committee is somewhat safe for the industry in that oil district congressmen are attracted to and stay on Public Works for reasons other than pollution legislation. But with the Senate Committee, chance appears to play an even greater role. When these recruitment processes are placed in the context of the Committee's operating style, we shall see that water pollution legislation in the Senate is often in the control of one individual.

Nevertheless, it is clear from our analysis here that committee recruitment has been an important factor in the superior success of the oil industry on the depletion allowance. Not that the industry has necessarily been unsuccessful in the pollution area, but committee recruitment is not as important a factor in obtaining superior success as on the depletion issue. The industry has not tried to influence recruitment to the Public Works Committees. Rather, it has been willing to deal with the membership that is "naturally" recruited to these committees to influence policy.

Committee Operations: The Water
Pollution Area

Of the four committees involved in the issues under study, the Senate Public Works Committee (and more specifically, its Subcommittee on Air and Water Pollution), is the only one where the recruitment practices do not have a direct effect on policy. What is important with regard to Senate Public Works (and, as we shall see shortly, with the other committees also) is the way it operates and the impact of its operations on policy and on interest group behavior.

With the Senate Public Works Committee and its subcommittees, the most important factor is the low level of attention given to them except from some senior members. Because Senators have numerous committee and subcommittee assignments, they are often forced to neglect certain committee duties. Duties on less important committees and subcommittees are likely to be the first discarded unless a Senator is the chairman or ranking minority member. Moreover, the vast majority of Senators are either a chairman or ranking minority member of a Senate committee or subcommittee and thus have bases from which to operate. With the exception of the more important Senate committees, Senators not in senior positions on a committee tend to focus their attention elsewhere. The one exception to this would be freshman Senators who may have no major positions. But they are likely to be the easiest for a committee or subcommittee chairman to deal with, given the norms of behavior for freshman Senators.[48]

The Muskie Subcommittee provides a clear illustration of this. Aside from Muskie and J. Caleb Boggs (Rep., Del.), the ranking minority member, few Senators gave it more than a passing glance. Since Muskie's Vice Presidential candidacy in 1968, some Republicans have tried to hinder his proposals. But many of those interviewed were quick to point out that Muskie still ran the show; Boggs and Howard Baker (Rep., Tenn.) were cooperative second and third bananas, and the addition of Senator Robert Dole of Kansas, then Chairman of the Republican National Committee, was the only factor liable to upset what Muskie decided to do.[49]

Thus, in the lower half of the Senate "caste system," the chairman and ranking minority member often have free reign in policy making. To the degree that a full committee grants a high level of autonomy to its subcommittees, the importance of the subcommittee is enhanced. The Muskie Subcommittee is a case in point. According to a member of Senator Randolph's staff, Muskie manages pollution bills all the way through, and the only time Randolph (Dem., W.Va.)—who is chairman of the full Committee and a member of the Subcommittee—interferes is when the legislation affects the coal industry.

As one might expect, changes in the chairmanship of a subcommittee can have big effects on the legislation the subcommittee produces. In the previous

chapter we discussed the effect on water pollution legislation, after Kerr's death, when the Muskie Subcommittee was created. The difference in policy was a direct reflection of the difference between Kerr's and Muskie's stance on the issue. True, Kerr was called the "King of the Senate." But even commoners like Muskie, when given control of an autonomous agenda-setting point in the legislative process, have major impact on policy. When respondents are asked why the Senate suddenly started to produce stronger water pollution bills than the House, the change from Kerr to Muskie is unanimously cited as a reason. As a member of the minority staff of the Muskie Subcommittee claimed: "It's primarily the people involved. It could change in the committee—right back the other way. You have a combination of an aggressive Senator plus good jurisdiction."

As with the case of changes in Treasury leadership in the tax area, the oil industry is well aware of the impact of Muskie on pollution legislation. One lobbyist claimed to admire Muskie because he had knowledge in the area and gave the industry a fair hearing even if he was in disagreement. He further commented that he was probably the only oil industry representative who held this evaluation. But this lobbyist and others agreed that Muskie is the focal point in the Senate, just as Kerr had been earlier and someone else would be later. Just as the Treasury provides a benchmark in the depletion allowance legislation, a Kerr or a Muskie can provide one in the pollution area. Moreover, a change of subcommittee chairman can do much to change the position of the Senate.

The House Public Works Committee and the way it operates present a dramatic contrast. Unlike the ferment that came with a change of chairman in the Senate, the House Committee presents a picture of stability. First, for most members of the House Committee, the Committee was the main focus of their attention. In 1969, ten members belonged to no other standing House Committee. Of the twenty holding other assignments on non-exclusive committees, only nine held more senior positions on those committees than on Public Works. Thus, for at least twenty-one of the members, other committee assignments did not have priority over their assignment on Public Works; for nine others, it could be argued that higher position on a non-exclusive committee may have held a greater portion of their attention than Public Works; and only three had the conflict of being on a committee of similar rank.[50] As opposed to the Senate Committee, most members of House Public Works are likely to see their assignment as a major part of their legislative responsibility. Moreover, even if it were not a focal point for its members, the sheer size of the Committee and its Rivers and Harbors Subcommittee (twenty-one members in 1969)—when compared to the Senate Committee and Muskie Subcommittee—increase the possibility of someone creating roadblocks for the chairman's desires. One oil lobbyist confirmed this in noting Muskie's ability to maintain control, "the House Committee is larger and has a number of members whose main interest is in the Committee and they've been there for some time." Blatnik has not had the same

free hand in writing water pollution legislation that Muskie has had. He has had to compromise with conservative factions in his own party's membership on the Committee to prevent defeat of legislation by a coalition of this faction with an active Republican minority. Murphy quotes one southern Democrat as saying about water pollution legislation, "We can scuttle the whole thing if we're together."[51] Blatnik himself will go down the list of Committee Democrats, stopping as he comes to a southerner and say, "He goes as far as he can with us." Then he will talk of the 1965 bill, when he had to compromise on the federal standards section to prevent the "conservative coalition" from defeating him. A bill, more like the one the Republicans wanted than the one Blatnik originally desired, was eventaully reported.[52]

The active participation of a large number of members makes the House Committee one where consensus must be met. With a highly partisan and active Republican minority, Blatnik was forced to compromise with oil district and southern Democrats on pollution legislation to report out any bill at all. Unlike the cooperation in the Senate between Muskie and Boggs, no such arrangement existed in the House between Blatnik and Cramer.

But is this really a picture of stability? After all, Public Works is a committee with high membership turnover. Doesn't this mean great change is possible as committee membership changes? Three qualifications are available to meet these reasonable inquiries. First, while there has been high membership turnover on the Committee, the leadership has been stable. Blatnik and Cramer have been on opposing sides on water pollution legislation throughout their tenure on the Committee. Now that Cramer is no longer in the House, William Harsha has assumed the position of ranking minority member. Although Harsha may not be as aggressive as Cramer, he still maintains the strong partisan outlook characteristic of the Committee. Additionally, as previously noted, a large number of Democrats have built up considerable tenure on the Committee. The norm of partisanship should remain even as many of the faces change.

Second, in many ways Public Works is the House in miniature. It reflects the House on measures of party regularity and support for a larger federal role.[53] To the degree that this continues to be true, we should expect the Committee to change no faster than the House does. The conservative coalition is likely to remain as an important factor in decision making. A vacancy is likely to be filled from states that have held the seat. There is no reason to expect the House leadership to suddenly become concerned over who gets on Public Works. Because of its size, elevation to chairmanship of the Committee or to a subcommittee chairmanship will take considerable time and is likely to produce a "well-socialized" member. Change even in this high turnover committee should be incremental.

Third, the norms of the Committee help preserve its stability on issues. Evidence I have gathered supports Murphy's claim that Public Works has "a decidedly constituency-oriented jurisdiction."[54] While water pollution legisla-

tion is seemingly different from other policy areas the Committee handles (because it is national in scope), the Committee operates in this area in a fashion similar to that used on single constituency legislation (which occupies most of its workload). That is, the Committee defers to Representatives from the affected constituencies. One oil district member of the Committee offered qualified support for this view:

I think you've got to realize that water pollution is different than those other areas. With roads or harbor projects the particular member has the best knowledge of his own area so it's natural to defer to him. It's not the same in water pollution. For example, I don't know much about drilling oil wells, but since I'm from _____ (his state) and the industry home offices are in my district they come to me or _____ (another oil district member) or _____ to present their case. They make several points and claim it will put them out of business—they're always saying that. But I consider whether their position's logical. And I'll make a case for those points. But the Committee doesn't really defer to me. They know because of my constituency that I'll be familiar with the industry case. It works that way with other people too; the members from the Southeast where they have some paper mills know the pulp industry position—some from big cities have steel. Members look to them to evaluate the impact of provisions.

Thus, although Blatnik is known as the House's man on water pollution legislation, he does not always exert control over legislation. In the Committee's executive session on the 1969 bill, Blatnik stepped back and let Jim Wright "run the show" on the oil pollution sections. The 1966 conference committee did not reach agreement until Wright had added the word "grossly" to the bill, making it unenforceable.

Given the bulk of legislation with which the Committee will continue to deal, it is likely that deference to constituency interest will continue to operate. When this is combined with the recruitment tendencies of the Committee, we find a picture of stability and protection for the oil industry. A change of leadership might make some marginal changes, but not in the same manner as a corresponding change on the Senate subcommittee.

The recent *Ralph Nader's Study Group Report on Water Pollution* criticizes John Blatnik for giving up the crusading fight for pollution reform that he made in the 1950s, for allowing "his Federal water pollution control family" to languish "for lack of effective House support."[55] The evidence presented here, however, indicates that little change has occurred in the pollution decision making apparatus in the House. Public Works was every bit as strong a "pro-industry" committee in the 1950s as it is now. Blatnik was forced to compromise then as now. The real change has come in the Senate, not in the House Committee. The benchmark on water pollution has moved for the Senate, but has been relatively stable in the House. The reason the Senate was the industry stronghold until the early 1960s was as much a function of Kerr's subcommittee as Blatnik's "aggressive drive."[56] Now Blatnik's drive appears,

according to the Nader study, to be in reverse. A more accurate interpretation is that Blatnik and the House Public Works Committee have not changed significantly.

What this means with regard to the oil industry position on water pollution legislation is that the House Committee provides the industry with a relatively stable benchmark from which to work. Depending on circumstances in the Senate Subcommittee, the House Committee's position will be either the best or the worst it can expect in the bargaining situation. The success of the industry position depends in large part on where the benchmarks are drawn. To cope with this the oil industry has regularly chosen to expend greater effort at the more favorable of the two committees. Prior to Kerr's death the Rivers and Harbors Subcommittee in the Senate was the place where industry interests were aggregated. Since the formation of the Muskie Subcommittee, the House Committee has been the major point of aggregation. Jennings reports how industry groups lined up on the Senate side, while conservation groups "channeled most of their efforts through Blatnik and the House."[57] After 1963 there was a marked turnaround. The oil industry, and industry groups in general, fired most of their shots on the House side. This amounted to more than just working through Wright and Edmondson. A major part of their impact was made through the House Committee's staff. Nick Gammelgard, chief pollution lobbyist for the American Petroleum Institute, and Richard J. Sullivan, Chief Counsel of the Committee, are in regular contact on all phases of pollution legislation affecting the industry. While there is nothing necessarily nefarious about this relationship, Sullivan is extremely well versed in the industry's position. During the 1969 executive session, when discussion turned to a provision for absolute liability for oil spills, Sullivan remarked that this section was totally unacceptable to the oil industry. The section was altered immediately.

This does not mean that the industry ignores the Senate Committee. One staff member of the Muskie Subcommittee saw the industry's lobbying resources as relatively unlimited:

If I were lobbying I'd focus on one or two or three people or points. Oil can focus on a series of people. Virtually every oil company has a vice president for each member of Congress. They can see everybody. . . .
Of course they come here (to the staff). They can't deal around us. They have to deal through us. Of course it only goes up to a point.

But another staff member whose personal position is closer to the industry offered a qualified picture of the lobbying activity:

I think the oil people fumble like hell. They're supposed to be an efficient lobby but they got clobbered. For example, a good friend of mine who handles work for oil came in on a Tuesday with amendments for Thursday's meeting. We thought he'd go to see the other members of the Committee before the meeting. But on Thursday when they came up, no one had seen them except the chairman (Randolph). Naturally they were defeated. In another case a bill went

all the way through with only one oil man seeing Senator Muskie. They assumed he was their adversary so didn't bother. Too often they rely on just three or four friends. The Senators don't have time enough to lobby each other any more.

This contrasts with the constant contact between the industry and the House staff. Moreover, it fits with the findings of Bauer, Pool, and Dexter from their study of trade legislation.[58] Whether this can be defended as a rational strategy on the part of the industry rests on one's view of the legislative process and resource limitations. Viewed as a serial process, it makes good sense to concentrate on the benchmark that can most easily be moved. From the viewpoint of the oil industry, the decision making institution that holds a position closest to its own is the one most likely to be sympathetic to its arguments and to move in the direction it desires. Success with any one institution may be the most significant impact that an interest group can have on the process. This is especially true when resources of a group are limited. But in the case of the oil industry individual decisions rather than resources are the limiting factor. Industry representatives feel that the Senate Subcommittee and its staff are too far away from the industry position for lobbying to be of much use. Industry officials will go through the formalities of discussing their position with some Committee members and staff, but without much hope of winning converts. They have people on the Committee who favor their position, according to a staff member. But they do not have enough of them who make the Committee a prime order of business to have much impact. Thus, the industry official who thought Muskie was a "reasonable" man had little hope of bringing him around to the industry point of view, especially given the nature of the Subcommittee staff. In talking of Leon Billings, Muskie's staff man on the Subcommittee, this official specified what he was up against:

Leon Billings is a very capable guy but he doesn't go along with the technical and economic feasibility problem. He says 'The industry has a lot of money. They should pay for the best equipment.' He's just not reasonable on this point. . . . In this sense the House staff is generally more reasonable. They know that a change is going to cost the industry money and that it's going to get passed on to the consumer.

What is clear from this analysis of the operation of the House Public Works Committee and the Muskie Subcommittee is that they are of equal importance in the making of water pollution policy. The difference between them lies in who has control over the legislation produced. On the Senate side, where the chairman's position is crucial, the industry's success is volatile in that a change in the chairman can turn the Senate around on the issue. Stability, deference to interested constituencies, and compromise are the keys to the House Committee. The industry has reacted to these circumstances by taking its case before the more favorable committee. When this was the Senate Subcommittee, the case

prior to 1963, the industry came away extremely well. In the 1963-69 period the industry had to rely on the House Committee, since the change from Kerr to Muskie on the Senate side meant the Senate position had jumped over the stable House position on water pollution legislation. The situation could, however, return to its pre-1963 status if the chairman who replaces Muskie is more favorable to industry.

The tax committees present a contrast to this situation. They are not of equal importance, and the oil industry's tactics are accordingly different. We now focus on them.

Committee Operations: The Tax Area

The importance of committee recruitment to the success of the oil industry's maintenance of the depletion allowance and how a change in recruitment is associated with the 1969 change in the allowance have been analyzed previously. Other factors, however, had an impact on the industry's long period of success and also help explain the greater success enjoyed by the industry in the Senate than in the House. In this regard, we have already examined the constituency composition of both the Ways and Means and Finance Committees.

However, additional conditions associated with the operations of these committees have been offered to explain interest group success. Most prominent among these is the "appeals court" theory. According to the "appeals court" argument, the reason the Senate treats the oil industry, as well as most other organized groups, more favorably than the House does is because the Senate always follows the House on revenue bills. This is part of a Constitutional requirement that gives the House priority in all bills relating to revenue. The contention is that the Senate acts as an appeals court for groups that fare poorly on the House side. (This same theory is applied to appropriations process in Congress.)[59] Some support for the theory is found among respondents I interviewed. A senior Ways and Means Republican believes that the "appeals court" process happened in all policy areas.

For those who don't think they did well in the House, they can continue in the Senate. Often the opposition, having won in the House, will not follow through in the Senate. You may be left just with groups trying to improve their position and even if some get their way the bill is weakened.

A concurrent opinion came from a staff member to a Finance Committee Democrat. He claimed that tax legislation is one area where no bill exists before Ways and Means writes it. According to him, "Groups don't get organized in time." By the time they get a fair notion of what Ways and Means is likely to do, there is insufficient time left to make an adequate case.

A similar opinion comes from the Treasury Department. Gerard M. Brannon formerly of the Office of Tax Analysis offered the following observation:

Tax legislation is different from other legislation because a bill pretty much spells out who is going to have to pay what. Each taxpayer can look at a tax bill and see how it is going to affect him. Thus, if there is a draft bill before the committee when the hearings begin, a lot of special interests will show up to complain that this or that section says diddley-do instead of diddley-dump.

Ways and Means, therefore, seldom has a draft bill until markup session is completed. This explains, at least in part, why pressure for special interest amendments focuses in the Senate Finance Committee rather than on Ways and Means.[60]

Other respondents, however, offered substantial disagreement with the appeals court argument. One Ways and Means Democrat was vehement in suggesting, "The special interests know what's happening here. The public doesn't." Later, he added, "Wilbur wants to keep things closed and quiet. He says if people find out what we're doing it'll affect the stock market. But special interests know what's going on anyway."

A Democratic Senator was just as strong in his disagreement with the appeals court idea. He saw the overall makeup of Finance and its leadership as the only real reason for the industry's greater success:

Long (Russell Long, Finance Chairman) is not just an appeals court. He would be for keeping depletion at 27½ no matter who acted first.

A *National Journal* article on the "money committees" equivocates between the two sides presented above. In one breath it calls the Finance Committee an "appeals court," since "few special interest groups see a final version of a revenue bill until it is out of Ways and Means," and in the next admits, "on the House side, pressure is applied surgically by an exclusive group of Washington tax lawyers. . . ."[61]

Additional information gathered and noted earlier tends to refute the "appeals court" notion, at least, as it concerns the depletion allowance. First, even without a bill, the line of debate on depletion is clearly set. Second, the industry monitors the activities of not only the Ways and Means Committee but also, prior to committee action, the Tax Legislative Council Office at the Treasury Department. It is actively involved from the beginning. Long before Ways and Means held hearings on the 1969 Tax Reform Bill, industry and Treasury representatives had met to discuss potential changes in the depletion allowance. Third, with the main exception of 1969, the industry position was upheld by Ways and Means. There was no adverse verdict to appeal to the Finance Committee. Finally, in 1969, had the process been reversed so that Ways and Means followed Finance, there is every reason to believe that the depletion allowance would have suffered a smaller cut, if any at all. Remember that the Finance Committee dealt with the House bill. If Finance had gone first, its members would have voted on cutting the allowance, not on restoring it. A tie vote would have meant defeat for a cut. Moreover, the industry position in

Finance would have been stronger on a vote to cut than on one to restore. The cut by Ways and Means provided a shield for Finance members who were marginal on the depletion issue.[62]

This does not mean that the process order on tax legislation has no impact on the lobbying of the oil industry as well as other organized groups. Again, the contrast with the water pollution issue is valuable. Instead of choosing the decision making institution whose position is closest to that of the industry and concentrating there, as with pollution legislation, the oil industry concentrated its efforts on Treasury and the Ways and Means Committee when it came to the depletion allowance. Clearly, the set ordering of the legislative process has an impact on the industry's efforts. The Finance Committee is an "appeals court" in the sense that the industry does not view it as an agenda-setting point. The reason the industry concentrates elsewhere, in spite of the fact that the Finance Committee is the decision making point most favorable to it, is that Finance deliberations are appeals in the sense that they come later in the process, after the boundaries of debate are set. The goal of the oil industry was to prevent depletion from becoming an issue by winning at an early stage in the process. Since there is often no draft bill before the Ways and Means bill, and since that becomes the working model for the rest of the process, it becomes vital for an interest group to win there or earlier. In the water pollution area the two committees each produce a bill somewhat independently of each other. They provide no distinct benchmark. In the tax area only one major benchmark is provided, the bill produced by Ways and Means. If there is a minor one, it is provided by the Treasury proposals.

Until 1969, the industry had either an executive that would not attempt to change the depletion allowances or a Ways and Means Committee that would protect it against most major changes offered by the Executive.[63] If Treasury or the President put changes in depletion allowances on the agenda, as in 1950, 1951, and 1963, the oil industry worked through Ways and Means to knock them out. When in 1969 some of the initiative for cutting depletion came from members of Ways and Means, the oil industry worked to withhold support of the Executive.[64] Previous groundwork by the industry proved successful. During the 1968 presidential campaign, the industry elicited Richard Nixon's verbal commitment for the preservation of depletion allowances.[65] Despite Treasury memoranda of July 17, 1969 and August 27, 1969 urging Nixon to support a "plowback" proposal for reforming the depletion allowance,[66] he refused on both occasions citing, as noted in Chapter 2, his earlier commitment.

Unfortunately for the oil industry, its success in holding back Nixon's support for a change was ineffective in stopping Ways and Means from voting a cut in depletion. Only in 1969 was Finance in a position to operate as an "appeals court." In this role the Finance Committee worked with the House passed-bill. And the tie vote of the "appeals court" meant a loss rather than a victory for the industry.[67]

On water pollution legislation, the House Committee bill became a benchmark for the House and the Senate Committee bill a benchmark for the Senate. The scope of debate was narrowed by the two committees. Interest groups work through the committee or subcommittee closest to their perspective. On tax legislation, efforts concentrate on Ways and Means and on Treasury. The former, however, is the more important. Even in 1969, when the President was closer to the oil industry position on the depletion allowance than was the Ways and Means Committee, the industry concentrated its efforts on Ways and Means. Perhaps this was because the President was already neutralized on the issue. (He would not support a change in depletion, but he did not *actively* oppose a change.) But a more accurate interpretation sees the Ways and Means Committee as the most important decision making point in the tax process as far as interest groups are concerned. Effectively organized interest groups do not wait for the Finance Committee hearings before they get into the act. Evidence indicates that the oil industry, among others, is in the process long before that. The industry knows what Treasury and Ways and Means are doing and hopes not to have to lobby beyond those decision making points. Winning where the agenda is set is more important. Beyond that the process becomes more public and private interests experience more difficulty.[68]

The Closed Rule

No consideration of the impact of rules, procedures, and processes on the strategy and success of an interest group in the area of tax legislation would be complete without discussing the "closed rule." It is one factor that clearly differentiates tax legislation from pollution, as well as most other, legislation. A "closed rule" is a rule provided by the Rules Committee of the House under which a bill is brought to the floor for debate and no amendments are allowed. The bill must be passed, defeated or recommitted to committee as it stands.

The most frequent granting of a "closed rule" comes on tax bills reported out of the Ways and Means Committee.[69] In the period under study no major tax bill has been debated on the floor under any conditions other than a closed rule. There is usually a debate and a vote on the floor on whether to accept the closed rule, but it always prevails. Once granted by the Rules Committee, a closed rule is not overturned on the House floor. In 1969, for example, the motion to move the previous question and adopt the closed rule on the Tax Reform Act carried 264-145.[70] Manley, in his study of Ways and Means, finds three major reasons offered for supporting the closed rule on tax legislation. First, tax legislation is too complex to be open to amendment from anyone on the House floor. Second, tax legislation, as well as other areas handled by Ways and Means, attracts interest group pressure. This could lead to a logroll of the House and a loss of revenue that Treasury could not sustain. (House members criticize the

Senate on this point. They see the Senate turning tax bills into so-called "Christmas tree" bills.) Third, the bills tend to be "major Administration measures," and proponents will vote for a closed rule to prevent opponents from weakening the legislation through amendments.[71]

Whether these are, in fact, the only reasons for the closed rule on tax legislation or just verbal rationalizations is not of major consequence here. The fact is that it does have an impact on interest group activity and on many areas of tax legislation—both complex and simple.

The most obvious impact of the closed rule is that it encourages interest groups to concentrate their efforts on the Ways and Means Committee. Since floor amendments are not possible, Ways and Means is the only place to have impact on the House bill. What Ways and Means produces usually winds up in conference. Although Manley finds disagreement among Committee members as to whether they feel any pressure from interest groups, at least with regard to depletion, the Committee has been the prime target of the oil industry, whose main weapon happens to be control of information and not pressure.[72] (We shall explore this question further in the next chapter.) Naturally, this further increases the importance of membership recruitment to the Committee.

The closed rule has other more subtle, but not less important consequences for interest group activity and effectiveness. It keeps one of the two visible stages in the tax process, the House floor, from playing a direct, significant role in the decision making process. The House can only vote a tax bill up or down. It cannot amend a bill.

Of course, to a degree the Committee must meet House expectations. But the point is that the House floor, which provides one of two opportunities (the other being the Senate floor) in Schattschneider's terms for massive "socialization of the conflict," becomes a place for approving not legislating. The Committee works in closed session, which is not easily conducive to expansion of conflict. It is hard to bring the general public into a fight when they are unaware a fight is going on. Private interests that monitor the process constantly can at least get a blow-by-blow account of what happened. As one staff member put it, talking about industry lobbying at the committee level:

Industry always has a spokesman (on the relevant committee). They are guaranteed access. Public interests have to rely on presence and the media.

When the Committee's decision is to keep something off the agenda, like not reporting out a section asking for decreases in depletion, the mass public is often totally unrepresented. The tax process only opens to conflict socialization when the bill finally reaches the Senate floor. By then it is often too late. Some palliatives may be added on the Senate floor, such as the Gore amendment to the 1969 bill that increased the personal income tax exemptions to $800.[73] This does not mean that the mass public necessarily gets involved at this stage, but it

is the first stage where the process is open enough to invite public involvement. Thus, Gore could obtain support by advocating an increase in the personal exemption. Such advocacy at an obscure point, however, generally elicits enmity from colleagues. Rather than being rewarded for espousing tax reform, one is more likely to be accused of "demagoguery with one's peers."

The closed rule helps to privatize conflict. Schattschneider claims that the reason private interests enter the political arena is because they cannot settle disputes among themselves. The closed rule helps keep the political arena for the writing of tax law a small one. True, conflict is to a degree socialized but there are controls such as the closed rule that limit the extent of socialization.

Finally, the closed rule was of specific help to the industry position on the depletion allowance. When combined with a controlled recruitment process to the Ways and Means Committee, the closed rule meant that House members were never given the option of voting to change the depletion allowance. One member of Ways and Means who feels that enough votes to cut the depletion allowance existed before 1969 claimed, "The closed rule is just the device by which they (Committee members) logroll. Put everything in a package and hold it together with the closed rule—the good and the bad." As long as no cut in the depletion allowance was in the package, the industry was safe. A strong pro-depletion membership on Ways and Means until 1969 assured this. To get at depletion the House would have had to defeat (or recommit with instruction) the entire bill. With a bill that contains numerous sections, which for most Congressmen are more desirable than getting a change in depletion, such activity was impossible.

One industry spokesman, while admitting that the closed rule had been a source of shelter from "Douglas" depletion amendments of a House variety, suggested that the tables were turned in 1969. He felt that, "in 1969, without the closed rule, it (27½ percent depletion for oil and gas) might have been logrolled back into the bill." Once the industry lost in the Committee, an appeal to the House floor was of little use. Without a closed rule, he felt, bargains could have been struck on the floor and depletion restored. In the sense that the closed rule reduced the number of possible decision points at which the industry could stop the change in depletion, the oil lobbyist was correct. In 1969 the industry might have been better off without a closed rule. But this view ignores the fact that even in 1969 it protected the industry from more serious cuts on the House floor and from attempts to pursue other tax provisions that benefit the oil industry. Intangible drilling expenses, for example, which are far more valuable to the industry than the change in depletion, were discussed in the Ways and Means executive session.[74] The issue was tabled there and was not brought up again for serious consideration in the House or Senate, although certain House members wanted to limit the application of intangible drilling expenses to exploratory wells only.

Another Factor: Incrementalism
as a Way of Life

Tax legislation is complicated, complex, voluminous, and intricate. The only generalities you get are people want taxes cut and want taxes even-handed. When Senator Douglas was on the Committee, he always talked of a provision being progressive or regressive. We can't write tax law on that basis. We don't get philosophical here.

—a Finance Committee Republican

In examining these two issue areas, one is struck by the attitude toward change enunciated by various decision makers. This attitude is epitomized in the above quotation and is best conceptualized in the literature on incrementalism in public policy.[75] A substantial segment of that literature concentrates on the budgetary process. My interviews affirm that incrementalism is not limited to this area. In both tax legislation and pollution legislation the attitude of decision makers is one of caution. Changes can only be achieved in small steps—each of which needs careful evaluation before another step can be taken. A Ways and Means Democrat who felt that the reforms of the 1969 bill were insufficient perceived the problems in getting further reform as follows: "Psychology is now that you've had a reform bill, you've got to slow down and wait awhile." Sometimes this "psychology" takes a slightly different form of expression and the argument for going slowly is couched in the language of a political broker. Thus a staff member of the House Public Works Committee observed that the Committee could not make any sudden changes, rather it must deal with "those who want to be left alone and those people gung ho who want to hit everyone over the head." His view was that the Committee's job was to balance these views. A senior Ways and Means Republican saw the tax legislative process in similar terms:

The Committee must balance the Treasury academicians who aren't always practical; the lawyers for the interest groups who are experts and practical but have a biased view; and political groups like labor, the Chamber of Commerce, and others who have philosophical points of view on tax legislation.

A third variation of this cautious attitude took the form of a rationalization— all you need are small changes to produce the desired policy consequences. This position is best exemplified in the statement of a senior northern Democrat on the House Public Works Committee. Known for his strong position on water pollution legislation, this gentleman surprised me when I asked about problems in enforcing the legislation. "Some," he claimed, "have too much of go-getter attitude on enforcement . . . like the speed cop who gives you a ticket for going one or two miles over the limit. . . . But if you know he's there, he doesn't have to give tickets for you to comply."

This manifestation of the incremental attitude is most critical in the behavior of the committees when they consider changes in depletion and water pollution legislation. With the exception of the Muskie Subcommittee, which will be dealt with shortly, the committees deal largely with short-term, narrow considerations when evaluating proposed changes. The Finance Committee considers the depletion allowance not in terms of its equity in the tax structure, but in terms of what it does for the oil industry in particular states. Correspondingly, the House Public Works Committee concentrates on the insurance problems of the oil industry, when oil pollution legislation is considered, and not on whether enforcement provisions are sufficient to deal with the problem. What this leads to is a phenomenon similar to the one Fenno finds common in observations of House Appropriation Committee critics. Just as they complain that the Appropriation Committee, in following its goals of budget cutting, has policy impact in failing to give programs adequate support, the committees on which attention has focused here skirt broad policy considerations in favor of incremental problem solving.

The Muskie Subcommittee is different, which is perhaps due more to its staff than to the attitude of its members. But for whatever reason, it has a clear goal. The people I spoke with were unanimous in their dissatisfaction with the inadequate water pollution legislation that they had managed to get enacted. This was true of both the Democratic and Republican sides of the Subcommittee.

The lobbyists were the ones most aware of the difference. Not only did the oil lobbyists view the Muskie Subcommittee as "unreasonable" and as trying to go too far too fast, but other lobbyists for "public" interest groups noted a vast difference in the approach to the water pollution problem taken by the two committees. A representative of the League of Women Voters suggested that the real difference between the two committees' handling of water pollution legislation was the way they looked at the problem. She suggested that Billings and Jorling, the majority and minority staff members for the Muskie Subcommittee respectively, were geared to the overall problem of stopping pollution and were in the "avant-garde" of the movement. On the other hand, she saw a member of the House Public Works Committee staff, as having an "interest in other things" in line with his machine politics background. A former Budget official, more graphic in describing the Committee staff's policy concerns, quipped about one individual, "I always keep my hand in my pocket when he's around."

One may yet wonder what difference this attitude of incrementalism makes regarding the success of the oil industry's position on these two issues. First, it combines with the industry's position of defending the status quo in each area to limit the likelihood of major change. Second, and perhaps more important, it steers the debate on the issues into areas where industry control of information and expertise becomes vital. This will be explored in more detail in a later

chapter, but let us briefly look at it with an example here. If the decision on the depletion allowance is based on the impact it will have on segments of the industry such as exploration and development of new sources, profitability of the industry, and risks in drilling, the supply of most of the information must come from the industry itself. The industry has been reluctant to supply this information even to the Treasury Department.[76] It becomes difficult, if not impossible, to build an adequate case opposed to maintenance of the depletion allowance on these criteria.

Such criteria should not be ignored when considering the depletion allowance, but when the discussion is confined to such a narrow perspective, the industry reaps a tremendous advantage.

Non-decisions

It is clear from our analysis that rules, procedures, and processes do have an impact on the effectiveness of the oil industry and the success of its policy positions in the legislative process. Not all of the factors we have examined have been formal structures that can be accurately delineated. There are some specifics, like the closed rule, but most are informal processes or ways of operating like incremental decision making. Naturally this latter group is more open to change. Thus, we should not be surprised if a new chairman of the Senate Subcommittee on Air and Water Pollution operated the Subcommittee differently from Muskie, just as Muskie operates differently than Kerr. But even some of the informal factors seem fairly stable. For example, there is little reason to believe that the House Public Works Committee will radically change in its approach to policy. It is large, has some stability in membership, and new members are quickly socialized to its style.[77] Informal processes often depend on one or a few individuals for their maintenance. Thus, a recruitment requirement that exists because Sam Rayburn is Speaker is likely to disappear when he dies, but an operating norm of a committee may continue long after the passing of individuals.

Certainly there are some limitations in the material covered here. Attention has been given to factors felt to be most important in affecting the success of the industry's position and on which I obtained information during interviews. Some factors about committee operations have been inadequately dealt with here.[78]

The factors we have observed may be separable, but they form a complicated web for explaining the industry's success. Why, for example, is recruitment to the relevant committees an important factor? Of course, we wanted to understand why there exists the disproportionate number of oil state Senators on the Finance Committee; to investigate the degree of policy relevance and industry activity in recruitment; and to explain how Ways and Means, with only 20 percent of its membership from oil districts, maintained such a strong

pro-depletion stance, at least until 1969. Of more value, however, is understanding why control at low visibility points early in the decision-making process is vital to industry success. Basically, the committee recruitment process is just another "mobilization of bias," rules of the game, which works to the benefit of certain groups at the expense of others. The rules of the game that we have examined operate to give the oil industry definite advantages in maintaining the status quo in the two issue areas. Some of these factors, like recruitment to the tax committees, have impact because of industry activity but others, like the decentralized decision making process, are there for reasons that extend beyond anything the industry does. The impact of these factors is to create a decision making environment in which non-decisions are a common form of policy output. The oil industry engages in activities that also encourage non-decisions. While the specific activity may vary, it is always geared to the same goal. Moreover, the oil industry is assisted by the nature of the rules and procedures of the legislative process.

Forms of Non-Decision

Most non-decision making has taken forms other than the most direct, "force . . . as the means of preventing demands for change."[79] By briefly reexamining some of the industry's activities, we can discern decisions that result in the absence of conflict.[80]

1. Membership on the Ways and Means or Finance Committee is a highly valued position. When membership is denied or one believes it to be denied unless one holds the proper position on the oil depletion allowance, the awarding of membership fits into the category of rewards and sanctions. This effectively prevented the demand of a change in depletion from garnering support in either of the two tax committees until the late 1960s.

2. Sanctions are not always so direct. Sometimes they involve only co-opting any opposition that exists. We have already seen the example of Herman Eberharter, whose attacks on the oil depletion allowance fell to the wayside with the inclusion of an increase in the allowance for coal. A more subtle example is provided in the implicit agreement in the House Public Works Committee to defer to members from affected constituencies.

3. Not all sources of non-decisions depend on activity of the oil industry. The fact that the legislative process is decentralized into many low visibility decision making points allows non-decisions to occur. As a staff member to the Muskie Subcommittee claimed, "A tough decision made at a low level of visibility is easier to make and maintain. It's also easier to make a weak decision." Since the game is played at a low level of visibility, weak decisions, which as far as the oil industry is concerned may be non-decision, are easier to make. When the audience is restricted, the private interests who monitor the low visibility points have an improved chance of winning.

In addition, something we did not discuss when looking at committee operations is that the committees with which we are concerned have a high level of acceptance on the floors of their respective chambers. Thus, a bill reported out of one of these committees is likely to provide an agenda for debate on an issue. When the House Public Works Committee reports a water pollution bill, there is rarely any discussion about what the Committee decided not to include. The Ways and Means Committee is shielded by the "closed rule." Only Finance has a great deal of difficulty on the floor. But when I asked why the Senate Floor repeatedly defeated motions to cut depletion by a three-to-one margin, and then in 1969 refused to restore a cut in depletion by a similar margin, the same answer came in one form or another from various respondents. Part of that answer was a claim to the effect that many Senators act to support the Finance Committee.

The important decisions in these issue areas are made at low visibility levels. Thus, a major part of the mobilization of bias is in placing important decisions at a level unseen by the general public, where socialization of conflict is low. Once a bill reaches the floor, conflict is low and often important provisions, having been defeated in committee, are not even on the agenda.[81]

Committees need not be points of low visibility but in general they are. While hearings may often be open, mark-up sessions are not. Until recently, the media gave only spot coverage to committee activity. And mark-ups, the most important part of a committee's work, were always done in executive session.

(A rules change at the beginning of the 93rd Congress provides for House committee sessions to be open unless the subject matter involves national security or is of a personal character, or unless the committee votes to close the session. This allows, but does not guarantee, open mark-ups.)

But the committee level is not the only place where the oil companies have worked for non-decisions, although it comes early in the process and by the industry's own admission is the most fruitful level of lobbying. Other low-visibility points in the process serve nearly as well. One, touched upon in our earlier analysis, is the conference committee. When the House and Senate pass different versions of the same bill, conferees must iron out these differences. Again working in closed session, it is easy for conferees to make a "weak decision" or no decision at all. Thus, Wright's supposedly harmless conference amendment to the 1966 water pollution bill made the enforcement section on oil spills meaningless. The 1969 bill nearly died in conference—after unanimous passage on the House and Senate floors. After the Finance Committee defeated an effort to restore the depletion allowance to 27½ percent for oil, Russell Long, anticipating the eventual conference with the House, requested that the Committee at least increase the House-passed 20 percent—which, as we shall discover later, kept the impact on the oil industry minimal.

The important point is that in neither issue area is the House or Senate floor the important decision point. The most public parts of the process are the places where the least important decisions are made, and only occasionally has the industry concentrated its efforts there.

Thus, the oil industry has at least three potential points of success. A process making these the most important stages for legislative decision making certainly favors organized private interests. As one oil industry official remarked, "once they (pieces of legislation) get to the floor, public support on the side opposed to industry is usually too strong."

This brings us full circle. Again, we are back to Froman's major premise. The Congressional process is a serial one. All the oil industry needed to do was win at one decision point to delay, defeat, or cripple proposed legislation in these issue areas. With the depletion allowance, when the Ways and Means Committee went against industry wishes in 1969, and Nixon and Treasury were neutralized, the industry still had the Finance Committee to assist it in recouping potential losses. As we shall see in Chapter 5, the consequences from the 1969 change were not as great as threatened. Similarly, after 1963 Kerr was no longer around to protect industry interests on Senate anti-pollution bills, but the House Committee remained a substantial obstacle to stronger anti-pollution legislation.

Given these factors, which reinforce the non-decision goal of the industry, the surprise is that any policy changes have occurred in these issue areas, and not that the changes were small and happened only after a long period of time.

Conclusion

It is quite clear from this chapter that rules, procedures, and processes do have an impact on the success the oil industry has on the depletion allowance and water pollution issues. Moreover, their impact on the depletion allowance seems more favorable to the industry than on the pollution issue. The closed rule and the control of recruitment to the tax committees have been the main reasons for the difference. In addition, some changes have taken place over the period under study and have corresponded with changes in policy. Recruitment requirements of the Ways and Means Committee and changes in the leadership of the Senate Subcommittee in charge of water pollution legislation are chief among these.

Yet despite these changes the industry would still seem to have had enough control over each of the issues to stop the legislation produced by 1969. After all, the changes in Ways and Means brought an anti-depletion minority to the Committee. And the switch from Kerr to Muskie on the Senate side still left the industry with a House Public Works Committee that was far from sympathetic to a strong federal water pollution control program. Despite these alterations, most of the process in each of the issue areas remained the same. It was still non-decision oriented. The changes that did take place were certainly necessary for policy change, but they were far from sufficient. Stoppage points still existed in the legislative process on each issue. Yet by 1969 the oil industry was unable to prevent reduction in the depletion allowance and the passage of an extensive oil pollution bill. An explanation of this must await the examination of other factors, which follows.

An auxiliary finding from the study of rules, procedures, and processes does offer some refinements to our understanding of interest group strategy. Dexter's claim about the tendency of interest groups to work with their friends is reinforced by the findings here. This is especially true in the study of water pollution legislation. Industry groups and conservation groups merely exchanged their points of concentration following Kerr's death. The former went from the Senate to the House Committee, and the latter from House to Senate. But the depletion issue provides an important qualification to this generalization. Although the Finance Committee was more favorable to the oil industry on depletion than Ways and Means, the industry concentrated its efforts on the House Committee. Prior to 1969 this could be explained simply by the fact that Ways and Means came first in the tax legislative process. With the industry preferring to win as early in that process as possible, it concentrated its efforts on Ways and Means. Once it won there, there was little need to exert much effort with the Finance Committee. But 1969 showed there was an additional reason for concentrating on Ways and Means. In a year when the industry did not prevent a cut in depletion in Ways and Means and also had to make its case on the Senate side, Ways and Means still received a disproportionate share of the industry's attention. In the tax legislative process, the committees are not equal in their impact on legislation. Ways and Means sets the agenda, and what it produces becomes the House bill. Unlike water pollution committees, Finance and Ways and Means are not independent partners in the decision making process. In effect, the industry does not have a choice on where to concentrate its activity on tax legislation. Thus, despite the more favorable surroundings provided by the Finance Committee, the main thrust of industry activity in 1969 continued to be Ways and Means.

Who Is On the Other Side?

The thrust of the two previous chapters has been directed at the relationship of the oil interest with the federal decision making apparatus. This focus, while a necessary one, has led us to ignore certain vital factors affecting the industry's success in the process. Thus, we have concentrated our attention on the strength and number of industry ties to decision makers, the advantage the industry obtains from being on the defensive side of issues, the industry's role in the committee recruitment process, and the overall impact that a low visibility, incremental, decentralized policy process has on the industry's success.

We derived certain advantages in analyzing each of these factors in interaction with the oil industry. In particular, this approach has lent a degree of clarity and simplicity to a complex set of relationships. But it is naive to think that the policy process can be well enough described in terms of these relationships. The policy environment in these issue areas is not necessarily the private preserve of the oil industry, the Congress, and the Executive. It is not just the oil industry trying to move the government (or keep it from moving) in certain policy directions. If that were the case, and it sometimes is, the industry would encounter little difficulty. As Schattschneider explained so well:

. . . the outcome of every conflict is determined by the extent to which the audience becomes involved in it. That is, the outcome of all conflict is determined by the scope of its contagion. The number of people involved in any conflict determines what happens; every change in the number of participants, every increase or reduction in the number of participants affects the result.[1]

A significant focus therefore involves who else gets into the fight in these issue areas and what factors contribute to or diminish the chances for contagion of the conflict. To do this we must direct attention to the oil industry's opposition in each of these issue areas. While the subject of opposition groups was not completely ignored in the earlier discussion of decision making visibility, it deserves more extensive consideration.

Clearly variability in the strength of opposition groups can affect the success of the oil industry's position. To illustrate, it may be helpful to think of this in terms of cross pressures. A Congressman presented with oil's position on an issue and with no significant opposition to that position is less likely to oppose the industry position than one aware of substantial, organized political opposition to the issue. In the latter situation cross pressures may make it more difficult for a Congressman to arrive at a position favorable to the oil industry. In fact, other

things being equal, he may take the side of the interest he perceives as stronger, or he may simply withdraw from active support of either. The particular resolution is not important here. What is important is to realize that success of any given group's policy position rests to some extent on the existence and strength of an opposition in addition to the factors already analyzed. Again we are interested in variations in the opposition between the two issue areas and over time.

We shall also consider the problems which groups, especially mass groups, face in having impact on the outcomes in each of these areas. Finally, our concern here is with how the nature of the opposition contributes to or alters what we have analyzed in previous chapters.

As in the other chapters, we shall first examine each issue area separately for the purpose of describing the nature of the opposition in each and its variability over time. The contrasts should then be obvious.

The Depletion Allowance—The Problems
of Organizing an Opposition

In his seminal 1957 article, Stanley Surrey elaborated the problem facing the public in influencing tax legislation. The question of who speaks for tax equity and tax fairness "is answered today largely in terms of only the Treasury Department. If the Department fails to respond, then tax fairness has no champion before the Congress."[2] One of Surrey's points in the article is that no one but Treasury opposes special interest provisions in the legislation. While there have been some recent changes in this relationship, a great deal of the oil industry's success in keeping depletion at 27½ percent until 1969 can be explained in terms of a lack of an organized opposition to the industry position. Surrey provides a more than adequate institutional explanation for this set of circumstances. And we shall regularly return to his explanatory factors in analyzing the changes which have occurred since 1957.

But certain limited insights can be gained into understanding why an organized opposition to the depletion allowance was lacking during most of this time period if the problem is discussed in light of the reasoning provided by one of the more enlightened critics of group theory, Mancur Olson, Jr. (Moreover, Olson provides the necessary conceptual logic to make comparisons with the case of opposition on the water pollution issue.)[3] Olson would argue that the lack of an organized opposition to the depletion allowance merely confirms the theories he presented in *The Logic of Collective Action.* That is "large or latent groups have no tendency voluntarily to act to further their common interests."[4] Thus, taxpayers (a latent group), an obvious source of opposition to special interest provisions like depletion, have lacked lobbies to exert pressure in testifying before committees, influencing committee assignments, and so forth. The costs of group action, according to Olson, have outweighed the gains.

Olson's lucid explanation, however, is too simple to deal with many of the complexities of the empirical political world. While some of the problems for those opposed to depletion stem from the cost of group organization, the lack of social pressure or incentives to enforce contributions to the group, and the irrationality of such voluntary contributions given the nature of collective goods (why should I contribute if I will get the collective good even if I do not?), an additional difficulty stems from the fact that the gain from a change in the depletion allowance is disaggregable. This significantly lowers the average interest of the individual in a change in the depletion allowance. While the change would take the form of a collective good for everyone who pays taxes, it would be divided among all the taxpayers. A Ways and Means Democrat put the problem simply, "You have a large group of taxpayers who have little to gain by reform and a few interests who have a lot to lose." A change in the depletion allowance has little impact on the taxes that any individual might pay. Therefore, the taxpayer has little reason to pay the costs of working to lower the depletion allowance. But to a major oil company the same change may be worth millions of dollars.

We should not limit ourselves to considering unorganized, large groups like taxpayers as the only source of opposition to the depletion allowance. As Olson notes, certain large economic groups organized for other purposes may lobby for collective goods.[5] In fact, during most of the period of study it is hard to find these groups lobbying to reduce the depletion allowance or for tax reform in general.

In 1950 and again in 1951, following the Truman tax messages (both of which called for lowering depletion for gas and oil to 15 percent), Ways and Means held lengthy hearings. After the 1950 message, 98 witnesses testified before the Committee either in favor of keeping oil and gas at 27½ percent or on raising rates for other substances. Only two witnesses, aside from Treasury Secretary John W. Snyder, favored lower depletion allowances. In neither case was depletion a main concern of the witness. Stanley Ruttenberg, testifying for the C.I.O., devoted little more than one sentence to the depletion issue.

As a matter of fact, we think that a reduction to 15 percent in the oil and gas depletion allowance is a step in the right direction.[6]

Similarly, Matthew Woll, on behalf of the A.F. of L., mentioned depletion in that small section of his statement on "closing loopholes."[7]

The 1951 story is just a repeat. There were 59 pro-depletion witnesses and 32 other statements submitted for the record that can be similarly classified.[8] Against this caravan of support came Snyder and his staff, Ruttenberg with a somewhat stronger statement, and the A.F. of L. with a minor comment. They were joined by representatives of the National Grange, claiming depletion "should bear a reasonable relationship to investment costs and to probably life of the producing resource. . . ." and the American Farm Bureau Federation

recommended "a careful study" of depletion allowances.[9] Again, depletion was not the axe these interests came to grind. It is not therefore surprising that the Ways and Means report does not mention anything about reduction in depletion allowances.[10]

Tom Field of Taxation with Representation, a new public interest tax lobby, claims that this situation is not unique to the Congressional side of the process.

It is a sad fact that almost no one now speaks for the general public when legislative hearings are held on tax matters. The situation in the case of administrative hearings is even worse; it has been many years since anyone without an axe to grind appeared at a hearing on a proposed Treasury regulation.[11]

Organized groups that might take the taxpayer's side are too busy with provisions directly affecting them. This, according to Surrey, combined with the complexity of the tax code, leaves Treasury standing alone.

The main reason is obvious. When the issue is a special provision for one group as against the tax-paying public as a whole, what pressure is there to speak for the public? Other legislation . . . brings forth strong and opposing pressure groups. But what pressure group fights against capital-gain treatment for employee stock options? Which group sees itself harmed by a 'Mayer amendment'? When the tax issues are at a major political level, as are tax rates on personal exemptions, then pressure groups, labor organizations, the Chamber of Commerce, the National Association of Manufacturers, and others, become concerned. But when the tax issues are technical, the pressure groups act only as proponents not as opponents . . .[12]

The natural question we are left with is how formidable an opponent is the Treasury Department. First, as is evident from the discussion of Treasury leadership in Chapter 2, the Department as often as not has been on the industry's side. Treasury under Humphrey, Anderson, and Connally was not a source of opposition to depletion. As recently as February 1972, Secretary Connally offered reassurance for the oil industry. As noted in Chapter 2, during testimony before the Senate Finance Committee on legislation to raise the debt ceiling, Connally was quoted as saying, "I don't consider the (oil) depletion allowance a loophole."[13]

But the depletion supporters still clearly perceive Treasury as the potential major source of their problems on the depletion issue. In spite of the Nixon Administration's failure to support a change in the depletion allowance in 1969, some industry officials blame Treasury and Nixon for the change. One lobbyist expressed this contradiction most vehemently: "We've had the biggest lobby in the United States opposed to us, the United States government. If one single thing defeated us in 1969 it was this Administration." But given the information we've gathered on Nixon's position, a refusal to accept even the "plowback" compromise, it is clear what this official really was alluding to—not that the

Administration at any point favored a change in depletion, but that it failed to use its full political resources to oppose the change. Treasury was a strong opponent only in the sense that it was a weak ally.

Clearly, there are times when the perception of Treasury as a major source of opposition to the depletion allowance is correct. Even when this is the case, as it was during most of the Kennedy-Johnson years, it does not mean the oil industry loses. As Surrey noted in 1957:

Thus, in the tax bouts that a congressman witnesses the Treasury is invariably in one corner of the ring. Assuming the Treasury decides to do battle, which is hardly a safe assumption at all times, it is the Treasury versus percentage depletion, the Treasury versus capital gains, the Treasury versus this constituent, the Treasury versus that private group. The effect on the congressman as a referee is inevitable. He simply cannot let every battle be won by the Treasury, and hence every so often he gives the victory to the sponsors of a special provision.[14]

Surrey also recognizes staff limitations at the Treasury Department. While an interest group can direct its attention to one particular facet of the tax code, Treasury must be geared up to deal with the broad range of interest claims. In 1969, according to the estimate of a former Treasury official, there were approximately sixteen full time tax lawyers working under Edwin Cohen and John Nolan on the Tax Reform Act. This particular official was the one responsible for work on the depletion allowance. And yet he was only able to devote "at most" one-half of his time to that issue. He felt that he could not successfully compete with the oil companies, which not only had more people working on depletion but which controlled most of the information on the industry's financial situation.

This leads to one final difficulty faced by both Treasury and other groups in influencing tax legislation. The tax code is so complex that an extremely high level of expertise is required for one to change any section of it. One Ways and Means member expressed this problem by saying, "I'll bet you can't even find the section of the tax code that requires you to pay income tax." Both Manley and Surrey find the complexity of tax law a major factor in analyzing how tax legislation is written.[15]

Given the level of complexity and the need for expertise in writing tax legislation, the question becomes who has the expertise. With some exceptions the technical expertise is employed by the interest groups that have a stake in the tax legislation. Moreover, experts are usually geared to the particular tax problems of the interests employing them. Thus, in the 1960s it was not surprising to find that most of the technical case on depletion was prepared by one man spending half his time. Even if taxpayers could overcome the organizational problems Olson describes, they would face the additional problem of obtaining the assistance of tax experts.

One tax lawyer with whom I spoke related a personal story that epitomizes

the problem. In 1969 he was asked to testify before Ways and Means on the proposed tax reform legislation as a private citizen-expert. He realized, first, that he could only cover a couple of items given constraints on his expertise and time, and on the Committee's patience.

One which I selected was the investment tax credit. I thought it tended to be inflationary. I expected my testimony would just become part of the record. But when I returned to Chicago, there was a story and my picture in *The Chicago Tribune*. I immediately went to my largest client and asked, 'Do you want to fire me?' He said why should he. Then I showed him the newspaper. He said, 'What the hell did you do that for?' After explaining my position to him he said he did not want me to feel that my public positions had to be in accord with his business interests. But he pointed out that the change I recommended would have cost the company $12 million last year.

Needless to say, this individual is reluctant to be a public advocate of changes in the tax code.

Over the years all these factors have worked to favor the oil industry in maintaining the depletion allowance. Members of Ways and Means rarely hear *expert* witnesses testify against depletion. And as Surrey claims "As a consequence, the Congressman does not see a dispute over a special provision as one between a particular group in the community and the rest of the taxpaying public. He sees it only as a contest between a private group and a government department."[16]

An Opposition Appears

The picture Surrey painted in 1957, which has been supplemented here, has undergone some significant changes. By 1969 and even earlier, a growing interest in tax reform and in changing the depletion allowance had developed. Nowhere is this more evident than in the 1969 Ways and Means Committee hearings. The Committee devoted two days to testimony on depletion allowances. During that time 23 witnesses testified in favor of the then current rates or in favor of increases in the net limitation above 50 percent. This group, as usual, included the representatives of both regional and national oil trade associations and other mineral associations. But unlike 1950 and 1951, when 33 and 28 Congressmen, respectively, appeared before Ways and Means in support of depletion allowance, only six did so in 1969.[17]

The opposition also differed. Seven witnesses testified for lowering or eliminating depletion allowances. Moreover, their statements were entirely devoted to the subject of taxation of natural resources. For the first time House members came before Ways and Means to make anti-depletion statements. Of the seven witnesses five were House members.[18] The other two witnesses were not ideological ones like those from the A.F. of L. and C.I.O. eighteen years

earlier, but were instead experts on taxation of minerals. One, Arthur Wright, a Yale University research economist, had been a consultant on the CONSAD Report. CONSAD, a consulting firm in Pittsburgh, was commissioned by Treasury during the Johnson Administration to study tax policies for treating natural resources. A major conclusion of the CONSAD Report was that tax policies resulted in additions to petroleum reserves of, at most, $150 million per year while having a tax cost of $1.3 billion.[19] Wright attacked tax policy toward natural resources on the bases of lack of tax fairness, waste of tax moneys, and administrative problems.[20] He did not argue that the oil companies were making too much money or that there was no need for a tax program to encourage exploration. Instead, he spoke to the Committee on more limited technical grounds substantiating his position with research that had been done in the field and not with polemic.

The other technical witness, Jerry M. Hamovit, had served in 1965 and 1966 as the Tax Legislative Counsel's specialist on natural resource matters. Again his testimony was technical, and he quickly divorced himself from strictly political considerations.

I do not appear before this committee to either condemn or defend the allowance for percentage depletion. I am testifying, however, because I believe that the present tax treatment of mineral production payments in practice distorts the percentage depletion deduction far beyond what Congress has traditionally understood were its limitations.[21]

The composition of witnesses was not the only, or even the major, factor differentiating the 1969 hearings from earlier ones. It is, however, the most visible indication of a change in the nature of the opposition. More important is the way Committee members treated industry witnesses. For the first time members asked probing and, at times, hostile questions of the industry representatives. The surprise is that only a few of the questions came from the strongest opponents of depletion. Gibbons and Corman asked some. Most came from members who occupy the middle to conservative end of the Committee— Byrnes, Mills, and Conable.[22] Byrnes, especially, questioned a group of industry witnesses at length on how Atlantic-Richfield (previously Atlantic Refinery) had a book income of $410 million between 1962 and 1968 and paid no Federal income taxes.[23] When an industry representative responded, "I can't speak conclusively to this particular subject on some of these corporate structures. I only know how it affects me personally." Byrnes addressed these officials with an implicit threat:[24]

It is incumbent upon representatives of the oil industry, who appear here in support of the continuation of the incentives that currently exist in the law, that they show us how these results occur and justify them.[25]

Later, Byrnes added:

Frankly, I no longer know what to write to my constituents. They write to me that they have read some place that an oil company made $450 million in a 4- or 5-year period, and that the company didn't pay a cent in taxes to Uncle Sam. They are correct in believing, of course, what they are paying is assessed at an increasingly heavy rate.[26]

It would be wrong to conclude that Byrnes opposed depletion and other tax provisions for the oil industry. He was merely in need of a way to justify them. This is far different from parading industry witnesses who are cordially received, make their statements, are thanked for their appearance and the information they have supplied to the Committee, and leave unquestioned and unchallenged. By 1969 there is clearly some active opposition to depletion allowances. But if Olson and Surrey are correct, why is there an opposition, how did it come to exist, and how does it affect the success of the industry position?

People with different perspectives on tax reform found similar explanations for the seemingly sudden opposition the oil industry faced in 1969. One, a Ways and Means Republican, saw it as the culmination of a decade long buildup. He felt that during time of crisis, such as a war for which "patriotic support" existed, or when taxes are low, it is hard to build public concern about tax inequities. During the late 1950s state and local taxes began to increase at tremendous rates, and while federal rates did not change, people were earning more and therefore moved into higher brackets. Thus, by the late 1960s, a ground swell of support in the public had built up for tax reform.

A former Treasury official sees this buildup being met by marginal reforms throughout the 1960s. He points to the fact that the 1962 Bill started by going after the most vulnerable provisions, entertainment expenses, foreign tax havens, etc. But he noted that there was "no point in starting off with oil. We don't want to lose the support of Kerr." In 1964 marginal changes in depletion were made by going after the property grouping provision. The real pressure, he claims, came in 1967 when Treasury had to commit itself to further tax reform in 1969 in order to get Congress to go along with the surtax. As with the Ways and Means member, he saw public pressure building up for reform.

Yet these same individuals will say that such pressures were not formally organized for tax reform. Perhaps the best source of explanation of this opposition is in terms of political decision makers responding to what they perceived to be demands emanating from potential interests or as Truman describes them "interests that are not at a particular point in time the basis of interactions among individuals, but that may become such."[27] By 1969 a number of Congressmen who either appeared as witnesses against the depletion allowance or were members of the Ways and Means Committee may have perceived the existence of a potential group among the taxpaying public interested in tax reform. Some felt this interest could be appealed to, and others

realized it had to be dealt with.[28] Thus, Byrnes does not really attack depletion during the hearings, but he is in fact asking the industry to provide him with a way to defend it to his constituents. The result may meet the demands of constituents, or it may, as we shall discuss later, just provide them with a certain level of symbolic reassurance.

A second indicator of the concern with this potential is the drop in the number and geographic distribution of the Congressmen who testified in favor of depletion. Each represents a major petroleum and mineral producing district. Of the five top oil counties in Texas, two are in the 21st district—George Mahon's; two in the 19th—O.C. Fisher's; and one in the 16th—Richard White's.[29] The districts represented by the four Texans who testified plus that of Omar Burleson, the 17th, are connected to form a solid bloc covering the state's richest producing areas of Central and West Texas.[30] Similarly, Wayne Aspinall's 4th district in Colorado was responsible for 70 percent of that state's production. The sixth Congressman to testify favorably, Page Belcher, represented Oklahoma's 1st district, which is not the leading oil production district in the state but whose major city, Tulsa, has been described as "one of those booming oil cities of the Southwest."[31] It borders the second largest petroleum producing county in the state. Thus, testimony came only from those Congressmen who have extremely large amounts of oil production in their districts—unlike 1950 and 1951, when Congressmen from relatively small production districts also testified. This may indicate the existence of cross pressure on some Congressmen with significant, though not enormous, oil production in their districts.[32]

The second cause for the growth of this opposition to depletion, even if the interest was just a potential one, is that in 1969 depletion and tax treatment of oil was just one part of a broader consideration of tax reform in general. It may be difficult to build public support for tax reform in general, but it is less difficult than to build it to change one specific special interest provision, like depletion. The change of one provision may not greatly affect the taxes of a given individual, but overall "tax reform" may be perceived as having significant impact on an individual's taxes. Tax reform, a broader issue, is therefore an aggregation of various provisions that individually have little impact on the average taxpayer but collectively have meaning for him. Moreover, as we shall discuss in the next chapter, to the degree that a given provision such as depletion takes on the symbolic attributes of tax reform in general, that provision may obtain the attention normally reserved, if it exists at all, for the broader issue.

The existence of a potential interest contributes to our understanding of the reasons for the existence of an opposition in 1969. What is more crucial to the analysis here is not just why an opposition exists, but what its impact is. One example of this impact is that in 1969 it became impossible to produce a bill that would be accepted as a tax reform bill without a change in the tax treatment of natural resources. Over the period of time we have examined,

depletion—and specifically depletion and the tax treatment of the oil industry—
became the symbol of tax inequity. Every individual I interviewed regarding the
depletion issue felt that the Tax Reform Act of 1969 would have lacked
credibility if a change in the depletion allowance had not been made. Even the
staunchest supporters of depletion considered this a major difficulty.

The second effect is even more important. The existence of a significant
opposition led for the first time to a clear split in the oil industry between the
major producers, most of whom have both domestic and foreign operations, and
the smaller producers, who have domestic operations only. To understand the
meaning of this split one must realize that tax provisions have differential effects
on various segments of the oil industry. Most small producers and wildcatters
never come close to receiving the full 27½ percent allowance. These smaller
independents have a lower ratio of net income to gross income than the majors
and therefore reach the 50 percent net limitation on depletion long before the
27½ percent of gross figure.[33] More important to the independents and
wildcatters is the provision allowing the expensing of intangible drilling costs.
Since an independent or wildcatter may sell off the development rights to a well,
intangibles allow him to do so and still keep a favorable tax situation. He can
expense the costs of drilling, rather than capitalizing and then depreciating them.
He need not wait for the well to produce, but can sell off those production
rights and immediately deduct the drilling costs from his income on the sale.
This gives him the capital position necessary for further exploration. The large
producer does not operate from such a tight capital situation. While expensing
rather than capitalizing drilling costs gives him some tax advantage, it is not as
critical as it is for the small producer.

When it appeared in 1969 that some adjustment would be made in the
provisions for treatment of oil, a split gradually developed between the majors
and the independents. Every effort was made to hide these differences at first.
At the Ways and Means hearing, Harold M. McClure, President of the Independent Petroleum Association of America (IPAA), led a delegation designed
to represent all facets of the oil industry. Aside from McClure, whose organization represents a broad range of producers (big to small),[34] the group also
included one representative from the American Petroleum Institute (API), the
lobbying arm of the major producers; one from drilling associations; and one
from an oil investment banking firm.[35] They covered the waterfront, defending
the existing tax treatment from their various perspectives. But after initial
testimony, Chairman Mills began to dig at the point of differential impact of the
two provisions, depletion and intangibles.

Chairman: Thinking about domestic production now, which is more controlling,
the 27½ percent or 50 percent of net income?

McClure: Speaking to that point, I would say in the majority of the production
it is more affected by the 27½ percent. Speaking more specifically as to a
marginal well, 50 percent of the net is the more controlling factor.

Chairman: I am talking about treating all costs of drilling as we now so-called tangible costs which are capitalized, spread over a period of time, and I assume recovered without regard to depletion.

Wallace Wilson of Continental Illinois National Bank: Yes, sir.

Chairman: ... You are saying that it would have a rather disastrous or at least unwholesome effect upon the smaller independent operator who would be the wildcatter.

Jack Abernathy of Big Chief Drilling Co.: It would perhaps require considerable payment of taxes during the very time when the producer is pretty well strapped to take care of this development program anyway. . . .[36]

During his questioning, Mills brought out the fact that depletion was the more important provision for the big operators and intangibles for the smaller concerns. The original unity of the oil position began to dissolve as each segment of the industry tried to protect its favorite provision from change. The split became more evident the next morning when McClure introduced a group of seven witnesses from various regional independent oil producing associations. While none of these favored lowering of depletion, each made it clear that few small producers received anywhere near 27½ percent. A sampling of their testimony will demonstrate the point:

Clinton Engstrand of the Cooperative Oil and Gas Association: Actually in Kansas in a study we conducted five years ago the average for the independent producers was 15 percent that we realized from depletion. . . .[37]

David Bell of the Ohio Oil and Gas Association: The Ohio Oil and Gas Association recently conducted a small informal and unscientific poll among a number of producers in order to determine what benefits they have derived from our Federal tax laws. As was expected, they were unanimous and emphatic in their views that if percentage depletion was reduced or if intangible drilling costs were eliminated, they would soon be forced out of business. Also after talking to these producers, we have surmised that the average effective percentage depletion realized in Ohio ranges from 12 to 18 percent. . . . Most production in Ohio is marginal, and we cannot realize anything near the full 27½ percent depletion. [38]

But Bell's testimony offered a clear contradiction. Although the 12 to 18 percent was an average figure, it made it quite clear that most Ohio producers not only would not be "forced out of business" by a reduction in percentage depletion but would not be affected at all by a reduction.

A subsequent witness, Ed Thompson of the Permian Basin Petroleum Association (western Texas and eastern New Mexico), a more profitable

production region and therefore more favorable to depletion, tried to smooth over the seeming contradiction by saying "it takes some 20's and 25's (even 27½'s) to bring home those 0's, 4's, 8's and 10's to the 16 percent average."[39] Yet he said "most fell below 20 percent and the highest was 25½ percent...."[40]

By the time the Finance Committee met the split had widened. An aide to Senator Fred Harris (Dem., Okla.) said that by the time things reached the Senate side some non-Washington based representatives of major companies were fighting to increase depletion to 33 percent, but others were more realistic and wanted to protect intangibles and keep depletion losses minimal. The aide referred specifically to the testimony of Mr. William Cleary, President of the Oklahoma Independent Petroleum Association. Cleary argued that Oklahoma independents managed only 21 percent because of the 50 percent net limitation.[41] These independents hoped that Finance, even if it went along with the House cut in percentage depletion, would raise the net limitation to help the smaller producer.[42] One group of independent producers, the Kansas Independent Oil and Gas Association, which claimed to have over 1,400 members, announced support for Proxmire's graduated depletion proposal. Proxmire made this a central theme of his testimony before the Finance Committee.[43]

Industry officials are reluctant to admit that there was a split in their ranks. An official of the A.P.I. couched his response in terms of appearances:

I think the industry was united on depletion. But for some reason we didn't give as unified a presentation as usual. . . . The independent's difference has always been there. It's not new.

Maybe the difference was not new, but for the first time, in 1969, the difference became important.

An I.P.A.A. official who has more of a mixed membership to represent was somewhat more candid:

I guess there was some of that (referring to the split), but we tried to prevent it. It's our feeling at I.P.A.A. that you're best off not to give on anything and not to admit that you will. Otherwise, you're like the farmer who's selling a goat and has a sign on it that says 'Goat $10 or less.' You know no one is going to pay $10 for that goat.

The trouble in 1969 was that the independents, afraid of losing their farms to the majors, were willing to sell the goat for $10 or less.

The entire situation is best summarized in a comment offered by Edwin Cohen, Assistant Secretary for Tax Policy. In a meeting with representatives of the major oil companies on August 26, 1969, Cohen explained what happened in Ways and Means. He felt that one group "hell bent to cut depletion" (no doubt Vanik, Gibbons, and Corman) and another group "hell bent to preserve intangibles" (Boggs and the independent producers) "stumbled over one another." The result was a cut in depletion to 20 percent, and intangibles were left untouched.

While there were no doubt other factors contributing to the industry split in 1969, the existence of an opposition to the depletion allowance was a necessary condition for the split. Without an opposition threatening the tax treatment of natural resources, there was no reason for the small producers to fear losing intangibles. Moreover, this split made it possible for some Congressmen to differentiate their position on depletion. Richard Bolling (Dem., Mo.) indicated that many Congressmen had favored depletion because it helped the independent, but they had no use for the major companies. Bolling, who at one time was Sam Rayburn's protegé in the House, felt that Rayburn's support for the depletion allowance was based on the populist notion of keeping the independents in business, and that he was actually opposed to the major oil companies.[44]

One should not interpret this analysis to mean that the appearance of an opposition has resulted in a complete reversal in the legislative treatment of natural resources. After all, in 1969 depletion was only cut to 22 percent and intangibles were untouched. The opposition generated was largely unorganized and had impact only because some members of Congress responded to what they viewed as a potential interest. One might suspect that a state of quiescence could easily return. To an extent, as shall be discussed in the next chapter, the 1969 change in depletion succeeded for precisely that reason.

What has happened instead has been an effort to institutionalize and organize potential tax reform interests. Three organizations have been established with this purpose in mind. The first, Taxation with Representation, is an organization designed to remove the organizational costs from the public's involvement with tax issues. The organization looks for individuals with expertise in tax matters to represent the public in lobbying for change. It is involving persons not directly tied to private interests in lobbying activities. Heavy reliance has been placed therefore on individuals from the academic community. The organization was formed in 1970, and as yet its effectiveness cannot be judged.

An even newer organization with somewhat similar if more political purpose, Ralph Nader's Tax Reform Research Group, was founded in 1971. Again it is difficult to judge exactly how well this organization will do. The greeting that Thomas Stanton, an attorney for the Nader group, received in testifying before the Senate Finance Committee indicates that it may face some rough going before it has any impact on tax matters. Stanton claims he was "thrown out" of the hearings. And while his description of the action is exaggerated, an examination of his testimony as reprinted in the *Washington Monthly*, bears out his perception. After a brief statement Senators Paul Fannin (Rep., Ariz.) and Wallace Bennett (Rep., Utah) accused Stanton of being a "biased" witness. They then proceeded to ignore his testimony, asked questions like: "What is your background and experience in business and government?",[45] and complained that their time was limited as in the following exchange:

Bennett: Is this going to be a lecture that will take us into this afternoon?

Stanton: You criticized my bias first; now I would like to present the facts.

Bennett: In how much depth? We have six more witnesses including that panel you talked about. It is 10 minutes to 12. How long are we going to have to sit here and listen to you?

Stanton: I am sorry you don't wish to listen to my presentation in detail, and I will submit something to the record.[46]

The third organization, Tax Analysts and Advocates, is also working to involve tax and economic specialists in the tax processes. It is providing an alternative to the usual career pattern and has been successful in recruiting former Treasury staff people. These three organizations have access to Treasury, which until recently was accessible only to corporation interests. Moreover, if they do nothing else, they improve the chances that some sort of expert testimony against special tax provisions will regularly be heard by the tax committees aside from that irregularly supplied by the Treasury Department.

Water Pollution—An Impotent Opposition

By comparison, the water pollution issue is one where, during the period under study, an organized opposition to industry positions on pollution control has always existed. Conservation groups constituted most of the opposition. Groups like the Izaak Walton League, the American Wildlife Association, and other conservation and sporting organizations regularly testified on behalf of new water pollution control legislation. In the late 1960s their efforts were joined by other organizations—most of them new—like Friends of the Earth and Nader groups. Also during this later period the League of Women Voters became active in lobbying for such legislation.

The impact of the constant existence of an organized force for pollution control legislation is that, at least on the surface, industry groups have been far less successful in maintaining their position than they have been in the tax area. This is evident from several perspectives. First, since 1948 the Public Works Committee of both the House and Senate have regularly reported out pollution control legislation. The opposition to the industry position has not allowed the non-decision type of committee action to occur, as it did with the depletion issue. Second, contrary to their position on floor votes on depletion, it is rare for legislators to vote against stronger pollution legislation. No matter how strong the industry group is in a given constituency, no Congressman or Senator can afford to vote "for pollution."

However, two questions remain to be answered. Why has there been active, organized public support for changes in water pollution laws while little support has existed for changes in depletion, and how has this opposition affected the oil industry's success and activity in this area?

Part of the answer to the first question comes directly from Olson. Clearly, conservation groups exist for non-political as well as political purposes. Therefore, their political organizational activities can be viewed as a by-product of their normal organizational operations.[47] The fixed cost of organization is already paid.

But this is hardly a sufficient differentiation of the level of opposition activity in the two issue areas. A second difference stems from the type of goods being considered. Unlike taxes, which are disaggregable, water pollution is not. A person who goes to a beach and finds it covered with an oil slick receives the whole impact of that condition. The slick is not apportioned to all the people who want to use the beach. Rather, each person receives the total impact. This means that the average interest of the individual in water pollution will tend to be larger than his interest in the depletion allowance. (Of course, even with disaggregation, if the total value of disaggregable good is large enough, the average interest may be greater concerning it than with some non-disaggregable good of small value.) While Olson's problem of collective goods and individual rationality still exists (why should I pay the costs to organize for stronger pollution enforcement when if someone else does, I'll still receive the benefits), the additional problem of the impact on each individual being very small does not exist.

A related distinction between the two issues is their degree of discernibility. An individual never really sees what the depletion allowance costs him in dollars and cents and in fact is likely to be unaware of it at most times. But pollution, especially oil pollution on bodies of water and beaches is highly visible. It is something the average person can see and recognize and something the television camera can film. Thus, not only is the impact of oil pollution likely to be undiluted but its occurrence is obvious. Although mystics might claim that it is easier to organize people around the abstract, in the political world concrete issues improve the prospects for mass organization and participation.[48]

This flows directly to another reason for the relative ease of organizing an opposition to the oil industry's position on water pollution legislation. Until recently, pollution, unlike taxation, has been seen as an issue of low complexity.[49] One industry environmental lobbyist lamented the difficulty of this problem, especially when compared to the relatively easy situation faced by his colleagues working on tax legislation:

On pollution anyone who reads a couple of articles thinks he's an expert. It's much easier for the public to get involved. They can see pollution. But with tax legislation, like depletion, most people don't know how it affects them.

Finally, there exist distributional answers for explaining both the relative ease of organization and increased effectiveness of the opposition on the water pollution as opposed to the depletion issue. People affected by water pollution live close to each other. Those affected by the Santa Barbara spill were neighbors and were likely to have more basis for personal interaction on that issue than

they would as taxpayers over the depletion allowance. More important, however, is that the people affected by pollution tend to live in the same jurisdiction as the polluters. Thus, a Congressman from an oil producing area finds constituent interests on both sides of the issue. He doesn't have just the oil companies claiming, "if you enact that law, you'll put us out of business." He also has citizens saying, "if you don't vote for stronger pollution control on oil, we'll put you out of office." These are simplifications of the situation, but they demonstrate the natural cross pressured situation likely to develop. Thus, the absence of floor amendments to the oil pollution sections of the 1966, 1968, or 1970 legislation is not surprising. Floor criticism is even limited. For example, Russell Long provided the only criticism of the Senate Bill in the 1969 floor debate of S.7. After Muskie and J. Caleb Boggs had presented their bill to the floor, Long claimed "I have come to the conclusion that the better solution . . . would be to adopt the provision of Section 17(e) of H.R. 4148 whereby each vessel . . . would be held liable for oil clean up costs of $100/gross ton or $10,000,000 whichever is lesser. . . ."[50] His statement drew support from Mike Gravel (Dem., Alaska), Randolph, who was also concerned that liability levels did not exceed insurable levels, and John Sherman Cooper (Rep., Ky.).[51] But support for S.7 was then offered from an unexpected source—George Murphy of California. Murphy represented a state severely affected by oil spills and was faced with a reelection fight in 1970. His statement centered on the fact that the bill "should be a particularly effective deterrent against the despoliation of our lakes, rivers, bays and, significantly too, in my state, California, our magnificent coastlines. . . ."[52] One wonders whether he had the previously "ultraconservative" Santa Barbara area in mind.[53] In any case, no amendment was offered, and S.7 passed the Senate 86-0.

With all these organizational advantages, one may wonder, and properly so, why these pollution control interests were stymied until 1970 in getting strong oil pollution legislation. I posed this problem to a number of respondents and received seemingly contradictory responses. Staff members on the Muskie Subcommittee complained of the inadequate job done, especially by conservation groups. One majority staffer was especially vehement in this regard:

Conservation groups stink. . . . The conservationists have never said anything. They come up here and have an orgasm in front of the committee and then run away. They never propose any language. Never even come back and ask how things are going.

This same respondent felt that the newer public interest lobbies suffered from different, though equally critical, problems:

They're limited by money and competence. Usually, they're spread too thin on issues. Most tend to be overimpressed by the fact that they're public interest groups. They seem to get involved in only the high visibility aspects and leave low visibility decisions for the private interests.
    Their interest is in the press coverage.

His counterpart on the minority staff had a similar but more moderate perspective:

If the President of Esso calls up, room is made in the schedule. Denny Hays (lobbyist for Friends of the Earth) is just another constituent. . . . They'll (the oil companies) have the language and argument all worked out. Combined with access that is vital. Failure of access and performance is often the case on the other side. Except for Nader, they just say, 'This is what you should do. . . .' Industry comes in and says here's a subparagraph that you should add.

This view was also shared by an environmental lobbyist for the League of Women Voters.[54] She compared the League's resources with that of the American Petroleum Institute:

They have money and money buys staff. I have one assistant and a secretary. The rest of my help are local volunteers. They know what goes on in their local area but they don't know about Washington. API has people around the country too; executives and employees. But they have the big Washington staffs too. I can't keep up with everything. Ten years ago I could, but not any more. There's just too much.

By comparison, a staff man for the House Public Works Committee was complimentary about the work of conservation groups. Given that these groups spend more time on the Senate side, this is indeed a strange perception. But after talking to this man for a while, I realized that the minimal efforts made by conservation groups fit in with the way he saw his job—find where the interests stand and then produce something that strikes a balance and can pass. Interest group activity, beyond the hearings stage, would have made his balancing act more difficult.[55]

Industry officials tended to "poor mouth" their position when faced by organized opposition. They were careful to separate out the newer "public interest" groups from the conservationists. One official complained that public interest groups "can do a thousand different things we can't do. If we even just tried them we'd be run out of town." When I probed by asking him about resources like money, experience, and expertise, he continued his tale of woe:

Their funds are tax free and they get a lot from the foundations. . . . I think they're much more expert than we are. I wish I had some of them. . . . The people who say that are a generation behind. Now I wouldn't have said that 10 years ago or even 5 years ago. The public interest groups are well led and well organized and well financed. Nader, I think he's great. And he's not the exception.

While the resources of environmental public interest groups have increased in recent times, it is hard to say that they are in the same ball park with the oil industry. All one has to do is look at the offices, staff sizes, and clerical help of API and a public interest organization to cast serious doubt on the industry claim. During the course of my interviews, I was regularly confronted by statements like the following:

We can't afford to take congressmen to lunch or entertain them.

—League of Women Voters Official

_____ (a public interest lobbyist) comes here to talk about a piece of legislation. She's done all the research on her own and when she leaves she's got to go back and type up her own report.

—Staff Member on Muskie Subcommittee

API is there all the time. Wherever the action is, they know what's going on.

—EPA Official comparing the oil industry
lobbying to public groups

But resource and access differences are not the only weaknesses of public interest groups. A major difficulty they face is lack of skill and experience. In a previous chapter it was noted how one Senate staff man thought that oil industry officials did an inadequate job. Unfortunately, public interest groups tend to make the same kinds of errors. But given their relatively weaker position to start with, these errors are more critical. An example will demonstrate this point. During an interview with an oil district member of the House Public Works Committee, I asked whether—given his substantial role in writing water pollution control legislation—he had much contact with environmental groups, or whether they just tended to work with the Senate Subcommittee. Since this man is a key person in writing the House bill and is not particularly favorably disposed to the oil industry, his response was of particular interest.

I've never thought about it before but as I do now I tend to find it true. Rarely do those groups come to me with arguments for changes in language or suggestions for provisions. I hope they don't do that because I'm _____ from _____(Name of city) and am just interested in oil. If that's the reason, it's pretty naive.... On consumer measures, I'm on Government Operations also; I see them all the time. Nader was here everytime I opened the door. He'd present his case and provide information on a variety of things. Some I'd support ... others I disagreed with. But they don't make specific suggestions to me on water pollution....

Whether out of ignorance or the feeling that this Congressman was on the industry's side, the failure to deal with him was a definite mistake by the environmental groups. I tend to believe it was a combination of both factors. Nader's people were willing to contact him on consumer matters, where he was perceived as being relatively sympathetic, and not on water pollution matters, where their perception of him was, no doubt, different. The Nader Report, *Water Wasteland*, which we previously noted as being politically naive, fails to mention this Congressman at all.[56]

The oil industry, given its resource superiority and the natural advantage that comes from playing defense, can better afford such voids in their efforts. The public interest groups cannot.

How do all of these factors affect the oil industry's capacity to respond? The major problem that organized opposing groups create for the oil industry is that they socialize the conflict. They make water pollution and control legislation more prominent. (Naturally, they receive a big boost from oil spills like the Torrey Canyon tanker spill off the British coast and the Union Oil blowout at one of its Santa Barbara offshore wells. These make the problem itself more visible.) This has reduced the strategies available to the industry. It cannot stop legislation from being reported out of the two committees or being passed on the floor. Its influence can be in weakening the provisions of the bills that are reported out, delaying House-Senate agreement in conference, and undermining administration of new control legislation. In the previous chapter we saw evidence of each of these strategies.

An organized opposition means that water pollution control bills eventually pass despite what industry groups do. During 1961-1970 four major water pollution laws were enacted. But because of the weaknesses of these groups, it means that each time a bill is passed work must start on a new piece of legislation to replace the weaknesses of the previous one. This means rebuilding support among the interests that have just succeeded in getting a piece of legislation passed.[57]

It is not meant to imply that the opposition has remained unchanged during the period under study. The activities of the newer environmental and public interest groups are far more extensive than those undertaken by the traditional conservation groups. They are beginning to monitor those parts of the policy process that previously received little attention. The one-word amendment that Jim Wright offered in the 1966 conference committee could no longer pass today. There was unanimous agreement among all respondents that the oil industry is in a far more difficult position now than it was in 1966. Part of this is due, according to respondents, to the dramatic oil spills that have occurred since then. Part is also due to the transfer of enforcement from state and local to federal levels and from Interior to EPA. But all also credited the new public interest environmental groups for taking a more active role in legislation and administration of pollution laws.

Summary and Conclusions

Understanding the nature and strength of the opposition goes a long way in helping us explain variability in the oil industry's success. While constituency ties of the industry were useful in describing the steady state situation and at which decision points the industry was most successful, they were not, because of their limited variability, effective for analyzing changes over time and from one issue area to another. Similarly rules, procedures, and processes were more useful mainly in discussing strategic problems and advantages for the industry lobbying activities.

Especially in the depletion area, the appearance of an opposition goes a long way in discriminating the events of 1969 from those of 1963, 1951, and 1950—not to mention the intervening years when depletion received no consideration. Clearly, a substantial part of the legend of the oil industry's power in American politics stems from its ability to maintain the depletion allowance. But with little non-governmental opposition, I question whether this was an accurate reason for ascribing power to the industry.[58] Similarly, the improvement in the quality of work and number of organized groups active on water pollution legislation is reflected in stronger legislation and enforcement in the oil pollution area.

The opposition does several things to make the industry's job more difficult. First, it provides a cross pressure on decision makers who have industry interests in their constituencies. Second, it can act to expand the conflict by making low visibility decisions more visible to mass publics through the use of various media sources. Third, it takes a potential issue and makes it actual. Fourth, in doing the other three it limits the number of decision points at which the oil industry can win and thus narrows the odds.

Naturally, the degree to which each of these four tasks can be performed depends on the resources, skill, and organization these groups can amass. On tax matters it means that the industry, whose position once was preserved by the Ways and Means Committee, by 1969 needed to wait for the industry-packed Finance Committee to prevent serious deterioration in depletion. On water pollution matters, things have gone beyond that. No longer is Robert Kerr in the Senate and chairing water pollution hearings, and no longer can the industry position be preserved by blatant amendments in conference. In both cases, the legislative changes have been incremental. But the fact remains that some changes have been made. In the pollution area, where the industry has always faced some opposition, and where the opposition has developed some organization and skill, the policy changes are more marked than in the tax area.

Moreover, the tasks facing these opposition groups differ from one issue to the other. Success of the opposition groups does not just indicate their level of resources, skills, and organization. It depends on certain steady state features, described in this chapter and the two previous ones, and on the efforts of the oil industry on each issue. Thus, for example, dealing with a Ways and Means Committee, where depletion was an important consideration in membership recruitment, poses for opposition groups a problem they are unlikely to face when trying to influence other policies.

One might claim that the level of opposition to the oil industry in these two issue areas really reflects conditions described in the two previous chapters and is therefore not an independent factor in understanding variability in the industry's success. The lack of opposition forces on the depletion issue prior to 1969 is a function of the corresponding lack of any receptive audience on the Ways and Means Committee. To a degree this contention is correct. When there is little

chance in winning, why fight? Even Treasury adopted this stance in 1963, when Secretary Dillon threw in the towel in response to Senator Williams' request.

But much of what has been displayed reflects the independence of this factor from those previously discussed. Surely, there are reasons for the variability in opposition. In this chapter we have examined reasons relating to problems in the organization of large groups and those which develop because of the nature of the issue—i.e., complexity, aggregability, and visibility of the issue.[59] The important point is, however, that variability in the level of opposition is clearly related to the level of the industry's success in each issue area and between the two issue areas.

In a broader sense, this chapter serves as a critique of Olson's theory on collective action. Olson's theory is useful but far from sufficient in making the types of discriminations now concerning political scientists in studying interest group behavior. Examples in the empirical political world show several of the theory's limitations. Most important is its failure to distinguish among the costs involved in different issues. Not all collective goods are the same. When the goods are not disaggregable, or when the average interest in the good remains high despite disaggregation, collective action on the part of large groups may not be nearly as irrational as Olson implies. Second, whether or not an opposition forms may be more a function of the visibility of the issue than whether collective action is rational or not. What is important in the study of the real political world is what affects the likelihood of a group forming and becoming active even if such behavior is irrational. To say such groups should not exist is of little help in a political world where they do exist. Third, because Olson's concentration on the size of groups, he fails to pay enough attention to varying costs of involvement in different issue areas. Clearly the differential expertise required in the two issues under study here indicates that costs of involvement in each are not the same, and high costs can deter "irrational political activity" that might otherwise occur.

Note that these comments are all headed in the same direction. The discriminations Olson makes are not terribly useful in explaining irrational political occurrence. If we merely know that something should not happen, how are we to explain and predict its happening when it does occur. To say irrational behavior is the cause provides little, if any, understanding.

Finally, Olson's work discards the importance of potential groups in the political process. Even when they fail to exist as formal organizations, whether for the reasons provided by Olson or David Truman, they may have important impact on the political process. Political decision makers may act as if such potential groups really do exist. Perhaps a Congressman will distinguish an organized group from some potential group, like taxpayers, in some regards. But as the depletion case shows, he knows that taxpayers, although unorganized and inactive as a group, may not be inactive as individuals on election day.

Most of the distinctions made in this chapter have gone beyond the confines

of Olson's theory. From him we get a grasp of why mass opposition groups have difficulty organizing in either issue area and why there has been organized opposition to the oil industry in the pollution area for a long time. But in explaining why opposition groups have more difficulty on the depletion issue or why they have increased in recent years in both issue areas, Olson is of marginal assistance. In fact, reliance on Olson leads one away from even asking these questions. Yet it is quite clear that the activities of irrational groups and potential groups do have an impact on the success of private, organized interest groups. And somewhat ironically, as we have seen, such irrational group activities on the water pollution issue are credited with having far more effect than the activities of "by-product" interest groups.

We now turn our focus to the one remaining variable in this analysis of interest group behavior, the policy variable, to examine what additionally can be explained about interest group success and strategy.

5 What is Left for Policy?

The variables we dealt with in the three previous chapters provide substantial explanations for variation in the success of the industry's position on each issue. Seemingly, this leaves little left for policy categories to explain and would lead us to question their value as independent variables for the study of the legislative process.

However, this viewpoint is shortsighted. Indeed, policy types do much to improve our understanding of variability in the oil industry's success and changes in its strategy over time. As will shortly become apparent, the depletion and pollution issues have moved through policy stages during the two decades under study. Not only are they not pure cases of symbolic and material policies, but also each has moved during the late 1960s from one policy level to another. And with this movement the industry's strategy and success has changed.

As before, we will examine each issue separately with concentration on the periods of change. Following this, certain generalizations applying to interest group activity and success as an issue moves from one policy level to another will be offered, and the interaction of policy type with the variables, previously analyzed, will be discussed. In this manner, I hope to show that the use of policy as an independent variable not only offers additional insight in explaining group activity in the legislative process but also leads us to understand the workings of other factors in a more systematic manner.

Depletion: Material Non-Decisions to Symbolic Politics

The legislative activity and the action that finally reduced the depletion allowance for oil and gas from 27½ percent to 22 percent represents the transfer of this issue from non-decision legislative phase to a symbolic politics phase. Prior to 1969 the legislative activity on depletion had consistently worked to prevent it from becoming a fullblown issue. True there were the Truman tax messages and the Kennedy 1963 proposal. In addition, Senators Humphrey, Douglas, Williams, and Proxmire made their regular depletion forays on the Senate floor. These attempts had little or no chance of immediate success.[1] They were serious only in that they kept the depletion allowance alive as an issue.

The oil industry was never placed in a position where it had to bargain away

part of the allowance. And all its efforts were designed to keep things that way. Attention was given to influencing recruitment to the tax committees and thus to smothering the issue at the earliest, least visible, points in the legislative process. As described in Chapter 3, the legislative process in general and certain facets of tax legislation in particular assisted in achieving this end. In addition, public concern about and awareness of the depletion issue was low. Low awareness and concern combined with low visibility created a reinforcing cycle. The low visibility of the issue made it difficult to arouse opposing interest and the lack of an opposing interest made it easy to keep decisions at low visibility points. Prior to 1968 the industry effectively stopped all attempts to alter seriously the depletion allowance. At no time did either of the tax writing committees or the Treasury undertake a detailed analysis of the impact of the allowance. Evidence of the continued need and success of the depletion allowance was supplied by industry witnesses. Their conclusions were rarely questioned.

Yet the consequences of these decisions not to change or even thoroughly investigate depletion were tangible. Depletion continued at 27½ percent and the oil companies continued to reduce their taxable income by substantial amounts.[2] Few symbolic reassurances were provided to the public because the public was largely unaware of depletion and what it meant. When issues are kept at low visibility levels and decisions are made at those levels, the size of the conflict, if indeed there is a conflict, can be controlled.[3]

By 1969 the situation had changed. First, the Ways and Means Committee was no longer as favorably disposed to the depletion allowance as before. Depletion had ceased to be a qualifying issue for membership on the Committee, and sufficient Committee turnover allowed a vocal anti-depletion minority to form on Ways and Means. Second, high taxes and tax reform became issues. The activities of the Ways and Means Committee attracted increased press attention. Third, public awareness of depletion—as a tax inequity—had grown. Several of the people I interviewed credited Paul Douglas and later William Proxmire with keeping the issue alive for years when there was no chance of changing depletion. Their doing so, the respondents felt, had educated others in the Congress and in the press to the inequities of the depletion allowance. Even if people did not understand the depletion allowance, many could identify it as a tax break for the oil industry. Fourth, the two preceding factors combined to make depletion the symbol of tax inequity.

These changes have all been discussed earlier, and we shall not further elaborate them here. What is important is their combined impact on the oil industry's effort to preserve its tax status. The year 1969 marked the end of the industry's ability to use non-decision tactics effectively to preserve its tax advantages. The evidence we have gathered indicates that the depletion issue went from a material, non-decision issue to an issue of symbolic politics.

Given this situation the oil industry faced the following problem. With the

demand for tax reform high in 1968 and 1969, almost any tax reform legislation that passed the Congress would have to make changes in the symbols of tax inequities. As we noted earlier, there was unanimous agreement among those we interviewed that a tax reform bill containing no change in depletion would not be accepted as a reform bill. One Ways and Means member was quoted as saying, "It (the oil depletion allowance) sticks out like a sore thumb."[4] How then could the oil industry preserve its tax position if a reform bill were passed? We found that in 1969 the oil industry, either consciously or unconsciously, adopted a strategy for dealing with the problem of the issue's movement out of the non-decision phase. It involved an attempt to quiet the arousal of support for tax reform through the use of symbolic changes without materially affecting the industry's tax position. The strategy had three components. One, realizing a cut in depletion is inevitable, never publicly admit that you will accept a cut and argue that any cut will have disastrous consequences. Two, privately work to keep the cut small. And three, protect all other features of your tax treatment from attack and try to improve on some.

The first of these has already been partially documented. Only after carefully examining what witnesses said during the hearings and receiving confirmation in interviews could I effectively question the sincerity of their continued support of depletion at 27½ percent. Most of the industry witnesses made strong pro-depletion appeals. Even the independent operators quoted in the previous chapter who claimed that they never realized anything like 27½ percent were vehement in their feelings that a cut in depletion would result in less exploration, a reduction in reserves of oil, increased prices for petroleum products, and a threat to our national defense.

These arguments were familiar ones the industry had made for better than two decades. In testimony before the Finance Committee in 1969, the same predictions of disaster were reiterated. Robert G. Dunlop, President of Sun Oil Company, appearing on behalf of the American Petroleum Institute, reminded the Committee of the dangers of an inadequate supply of domestic oil.

A reduction in the depletion allowance would either result in higher product prices or in increased dependence on less secure foreign crude. Neither alternative is desirable.

Higher prices for gasoline and heating oil, for example, would fall more heavily on the nation's lower income groups since they spend a much larger proportion of their income on such necessities.

If prices do not increase, the reduced depletion allowance would result in reduced investment in domestic exploration and development, and increased reliance on foreign oil.

. . . In accepting this alternative we would be assuming a long-run risk that cannot be measured in monetary terms. We could be drawn into a conflict in the Middle East, in an attempt to insure stability. A substantially increased U.S. role in the Middle East could well lead to a direct confrontation between the two nuclear superpowers.

. . . An adequate domestic petroleum supply is a critical need for the security of the United States.[5]

These sentiments were echoed when debate on depletion reached the Senate floor. While most pro-depletion Senators dealt more with the direct merits of depletion than Mr. Dunlop, they usually managed somewhere in their speeches to allude to the disastrous effects of any change. Senator Ted Stevens (Rep., Alaska), midway through a carefully reasoned speech on the problems of maintaining adequate oil reserves and the need to encourage exploration, especially in Alaska, raised the issue of risks to U.S. foreign policy:

If just half that supply is jeopardized either by conflict in the Middle East or by a civil war in Latin America—certainly very possible military action would protect our security. Is this what the proponents of the cut in the depletion allowance really want? Our flexible foreign policy concepts would be seriously affected insofar as oil producing nations were concerned. Another Vietnam-type situation is a definite possibility. Exploration of any further American oil producing properties in foreign countries could only produce an undesirable and precipitous action on our part if we do not retain the present balance in our domestic producing capability *vis-à-vis* foreign imports. Clearly, this is not a desirable position in which to place our Nation.[6]

Yet, as early as the House hearings, it had been clear to many industry officials that some change in the depletion allowance would be made. And while they never publicly admitted their acceptance of some cut, it was clear from activities which occurred after the Ways and Means hearings that some small decrease would be acceptable to the industry. During the Committee's executive session four motions were offered to cut percentage depletion. Two of the four had industry support. One, by George Bush, the most articulate spokesman for the industry position, would have cut depletion for oil and gas to 23 percent, and the other, by Rogers Morton, would have made a 20 percent reduction in all depletion rates (reducing oil to approximately 22 percent). In Table 2-5 one can seen that these motions received backing from the strong depletion allowance supporters and were opposed by those favoring more severe cuts in depletion. This action, while consistent with a desire to minimize the cut in depletion, is a far cry from a position of all-out industry opposition to any cut in depletion. The Bush and Morton motions were designed to keep the cut small and were made in hope of frustrating efforts by other Ways and Means members, like Sam Gibbons, at making more drastic cuts. The Morton motion nearly succeeded. It received support from Byrnes and Conable, both of whom later voted for the Boggs motion to cut depletion to 20 percent. And Mills, Watts, and Ullman passed at first. Only after all the other members of the Committee voted did they cast the deciding votes on Morton's motion.[7]

One respondent also viewed the Boggs' motion as the same sort of effort.[8] But the industry felt that a large cut in depletion still had to be opposed. As one

industry official put it " . . . it was apparent there would be some reduction in percentage depletion. But I don't know where he got the 20 percent figure from."

Once Ways and Means voted to cut depletion for oil to 20 percent, the industry was forced to work to bring that cut back to a more acceptable level. The efforts in the Senate Finance Committee were in part geared to establishing a bargaining position from which to salvage part of the House cut in depletion. This is clear in Russell Long's request that the Committee, after the tie vote on the question of restoring depletion to 27½ percent, raise the House figure to 23 percent "to give us something to bargain with."

At the same time the industry used the depletion issue to bargain for preserving other tax provisions and even tried to improve its position on some. (This strategy was especially important to the independent producers for whom these other provisions were particularly crucial.) Immediately after Ways and Means voted to cut depletion, Byrnes raised questions about the expensing of drilling intangibles. He proposed that the first well on each lease is "exploratory" and should be expensed, but that each succeeding well should be considered "development" with intangibles capitalized and amortized over a five year period. Hale Boggs, whose depletion motion was evidently designed to prevent further changes in the industry tax structure, quickly responded: "We've done enough in one industry."[9] By a close voice vote, the Byrnes' motion was then rejected.

Another example of this strategy occurred during the Senate floor debate. One senator, Robert Dole (Rep., Kansas), was so concerned with another provision affecting intangibles that he admitted that depletion was a lost issue.[10] While speaking in support of the Ellender amendment to reinstate depletion at 27½ percent, Dole remarked:

I believe most of those in the oil industry realize that if we are going to have tax reform, it will be painful to those who are reformed. I believe most of those in the oil industry in my state would probably agree, though reluctantly, that some change in the depletion allowance will occur before this bill is finally enacted.[11]

Dole, whose concern was probably geared more to the independent operators, who are dominant in the Kansas oil industry, went on to argue against the 5 percent minimum tax on intangibles in the Finance Committee bill. Both the Nixon Administration and the Ways and Means Committee had included intangibles as a limited tax preference (LTP) item. And while this was only one of five LTP items, and the long range impact of this change was estimated by Treasury at $85 million per year, industry people feared it would attract investment elsewhere.[12] It remained in the tax bill when the Finance Committee substituted a 5 percent minimum tax provision for LTP.[13] Dole claimed that this provision:

is much more harmful to the oil industry, or at least I am so informed, than would be keeping the depletion allowance at 23 percent. In fact, if there were some assurance that we could keep the Senate bill at 23 percent on the depletion allowance, and strike out the 5-percent additional tax on special tax privileges which include the depletion allowance, I think most people in the oil industry, while they would not be satisfied, would accept what we did.[14]

The Senate accepted the minimum tax provision. In conference, however, where most of the Senate minimum tax provision was accepted by the House conferees, intangible drilling costs were dropped as a preference item.

In addition to these two attempts to use the cut in depletion as a bargaining item to save other provisions from attack, there was an effort to improve the industry's position on one provision. This involved raising the percentage net limitation on depletion above 50 percent.[15] Hale Boggs first attempted this in the Ways and Means executive session as a trade-off for lowering depletion to 20 percent. He first proposed increasing the net limitation to 70 percent. Treasury opposed this as costing more than would be gained by the corresponding decrease in depletion. Boggs then tried 60 percent, but even there could not keep the support of anti-depletion allowance members interested in a more drastic cut.

In the Finance Committee, Russell Long successfully added a provision to increase the net limitation, and the circumstances surrounding its addition give further credence to the notion that depletion was used as a tradeoff. After Long had depletion for oil partially restored (up to 23 percent), he turned to Tom Vail, Chief Counsel for the Committee, and asked, "Isn't there something we can do for the little guy?" Vail, according to a witness to the proceedings, was unprepared for the question and first suggested the Proxmire direct drilling proposal.[16] This proposal was designed to give further tax benefits for exploratory wells while eliminating the expensing of intangibles for development wells. Long quickly rejected this and asked if there were anything else. Vail responded with the idea of changing the percentage net limitation to 65 percent. Debate then centered on who is a "little guy." Long suggested it was an operator with less than $5 million net income. Someone pointed out that this would mean an operation with a $25 to $30 million gross income. Long then responded that he meant gross, not net. At this point, Senator John Williams decided some limit had to be put to this provision and is quoted as saying:

I can see that I surely went into the wrong business. In the chicken feed business anyone who grosses $5 million is a big guy.

Long immediately withdrew to a $3 million limit.[17] This provision was subsequently dropped in conference.

The net result of all the decisions on the variety of provisions affecting the oil industry was the following: (1) the percentage depletion on oil and gas was reduced from 27½ percent to 22 percent; and (2) "intangibles remain . . . fully

deductible in the year incurred and not subject to any recapture rule on subsequent sale of the underlying property."[18]

Evidence for the success of this strategy is more fully apparent when we examine the effect this change in depletion has had on the taxes of the oil industry.[19] After the House bill was passed, Treasury estimated the reduction in percentage depletion to 20 percent would provide Treasury with an additional $400 million in taxes.[20] Yet in 1970, the major U.S. oil companies paid only $185 million more in federal tax than they had in 1969. Part of the discrepancy is due to the fact that the estimate was based on the 20 percent rate of the House bill, not the 22 percent rate as in the final bill. A later figure based on a reduction to 22 percent estimated increased revenues of only $235 million (using the Senate-passed 23 percent figure, the revenue estimate drops to $150 million). Part is also due to the fact that the majors are not the only producers. (But they were the ones who should have been affected by the change in depletion.)[21] However, a major adjustment in the other direction offsets much of this. In 1970, the major oil companies had increased revenues of over $800 million over 1969. Therefore, we should expect that they would have paid more federal taxes. And if we recalculate their 1970 federal taxes based on the 1969 rate (8.7 percent in 1970 vs. 7.3 percent in 1969) and subtract from that the actual 1969 federal taxes, we find the major companies paid only about $60 million more in taxes when we control for net income.[22]

Upon further investigation we find other factors that dilute the impact of this change in the depletion allowance. The number of dollars an oil company receives is a function of the depletion rate and the company's gross income. Vertically integrated companies can use the pricing system for crude oil to adjust the gross income figure. Under Internal Revenue Service regulations an oil company producing its own crude can price that oil in terms of the prices it pays for crude produced by other companies. Since a large company buys relatively little oil from outside producers, it will often benefit after taxes by offering to pay higher crude prices. The increased price it pays for crude to outsiders becomes the price of its own oil at the well head. With this well head price used to compute depletion, the tax benefits from increased depletion more than offset the increased price paid to outsiders. The result of higher crude pricing is a transfer of profits from the refining end to the production end. The latter is subject to depletion while the former is not.[23]

In February 1969, when the Ways and Means Committee began considering tax reform legislation, Texaco announced that it would offer to pay 20¢ more for a barrel of crude oil. Some companies followed, but surprisingly several majors did not. The result was a price increase of 15¢ a barrel when the market settled.[24] Then in November 1970 Gulf Oil announced a 25¢ a barrel increase in the price of crude oil (what Gulf would offer to pay). During the two weeks that followed most major oil companies followed suit.[25] Considering that these were the first nationwide price increases for crude since 1957, and considering the

impact of a price increase on depletion, it seems likely that the move was tied to the change in depletion rates. The Texaco move may have failed to gain full industry support because it was premature. The Gulf increase is more clearly linked to the depletion change. *The New York Times* article on the Gulf price hike claimed:

Some industry observers contended that the move was an attempt by Gulf to recoup the earnings it lost when the oil depletion allowance was cut earlier this year from 27½ percent to 22 percent. 'A 25 cent increase will just about do it,' one analyst commented.[26]

What these observers did not explain is that the recouping was to be done by increasing the value of oil at the well head and thus increasing the amount of depletion received.

One tax expert, who believes the cut in depletion had practically no impact on industry taxes, argues that while major companies may not be able to claim as much for depletion, they can still avoid paying increased taxes. He contends that companies with foreign operations fail to use their foreign tax credits to the full extent available and that they reserve them for use in succeeding years. All the change in the depletion allowance means is that those companies can use more of the tax credit in place of the allowance.[27]

With the exception of industry representatives, all respondents felt that the impact of the depletion change in 1969 was minimal on the oil industry—even without the price increase.[28] If there was any impact it was not on the industry but on the public. As one Washington tax lawyer put it, "Public opinion thought there had been a considerable change, but really not much was lost." He felt that the industry had protected intangibles and had kept the depletion change within bounds. A staff member to an oil state Senator was more cynical in his evaluation. He believed the reason for the change in depletion and the preservation of intangibles was due to the symbolic nature of depletion:

Senators know where the meat in the coconut is. Depletion was a smokescreen to the extent that it gave Senators a chance to show they were for reform of the industry's taxes without having a major effect.

We would be wrong to conclude that the 1969 action was purely a victory for the oil industry. In some senses it was definitely a victory. Changes in the tax status of the industry were marginal. Most important provisions went untouched. In addition, the symbolic politics activity appears to have bought the industry time. While Ways and Means members disagreed over the success of the 1969 bill, they all felt it would be a while before they would again consider major tax reform. One Committee Democrat, in noting the wait-and-see attitude, found another problem resulting from the 1969 action:

Psychologically, now that you've had a reform bill, you've got to slow down and wait a while. In addition, there probably aren't going to be any more horror

stories. The run of the mill audience wasn't shocked by depletion but by the stories of people with high incomes who weren't paying any taxes.

The member was referring to cases cited in Treasury tax studies showing that 21 millionaires paid no federal income taxes in 1968. He felt that the changes made by the 1969 bill reduced the possibility of this occurring without making substantial changes in the tax law.

Despite the increased clamor for tax reform since 1969 and the Ways and Means hearings on tax reform in early 1973, no new tax reform bill legislation has been presented to the Congress. The Committee, instead, put tax reform aside because of the urgency of new trade legislation. Even if Ways and Means returns to tax reform matters in 1974, and if the Congress does pass new legislation, the impact of any new provisions would not be felt at least until 1975. And those are some very big "ifs." Since tax reform will inevitably increase someone's taxes, Congress is often reluctant to pass a major reform bill in an election year. Thus, at least six years will have gone by before a new tax reform bill appears.

In a broader sense, however, industry officials are correct in the feeling that they took a beating in 1969. They were forced to change their strategy for dealing with the depletion issue. No longer could they prevent it from being an issue. For the first time, the industry had to appeal the Ways and Means decision and try to win at a later decision point. For the first time the scope of the conflict eluded the industry's control. The use of a symbolic politics strategy is far less desirable than a non-decision one. The former works to privatize the conflict while the latter admits the conflict has been socialized, and at times encourages its further contagion. The hope is that the granting of symbolic reassurances to the participants and observers will bring about quiescence and a return to non-decision, privatized politics.

But symbolic reassurances do not always bring about quiescence. As we shall see when we examine the changes which are occurring in the water pollution area, symbolic reassurances can act as rewards to weakly organized and unorganized interests. Rather than undercutting the opposition such rewards may, in fact, encourage new political demands by providing positive reinforcement to those involved.[29] Privatized politics does not run the risk of doing this. Symbolic politics does.

Naturally, factors such as the level of organization of the mass interest and the possibility of clearcut evaluation of the reward affect the chances for quiescence. But even in the tax area, where public interests are poorly organized and where accurate judgments of the impact of changes are hard to make, even for the experts, the potential for continued arousal after symbolic reassurances exists. One oil lobbyist saw this as a major problem:

In 1969 some said 'We'll put this matter to rest.' But when the country is short of money we'll have another go at it. The allowance is still there.

Indications are that the 1969 Tax Reform Bill did not bring about quiescence. Instead it promoted an organized opposition in the form of the Nader Tax Group, Taxation with Representation, and Tax Analysts, whose members, encouraged by the 1969 activity, believe that they can have substantial impact next time around. Moreover, tax levels combined with general economic conditions continue to make tax reform a salient issue.[30]

I do not claim that the industry's resorting to symbolic politics led to this situation. In fact, as is manifest in the previous chapter, the reverse is the case. The existence of an opposition led to the adoption of a symbolic politics strategy. But the need to use this strategy in 1969 is indicative of a significant deterioration in the industry's position as the issue moves away from the non-decision policy phase. While the outcome of the next legislative effort at tax reform cannot be accurately predicted, it is likely that the oil industry will not be able to return to a non-decision state and will again be faced with accepting some incremental policy changes as a form of symbolic reassurance. The question then becomes how many symbolic rewards can be offered before they start to add up to material policy changes.

Since 1969, the oil industry has made some efforts to improve its position for the next go-round. The unsuccessful effort to place freshman Senator Lloyd Bentsen on the Finance Committee was designed to shore up the industry's position there.[31] And although industry officials deny that they took an active role in the context between Joe Waggonner (Dem., La.) and Don Fraser (Dem., Minn.) for the Ways and Means vacancy created when Hale Boggs became majority leader, their preference was well known. It is unlikely, however, that the industry will ever have the control over recruitment to these committees that it possessed prior to 1961.

Other oil industry activities have involved attempts to create a competing symbol based on the existence of a so-called energy crisis. Since this new line of activity is being used on both the tax and pollution issues, I shall reserve further comment on it until I discuss the consequences of issue movement in the oil pollution area.

Water Pollution Legislation: Symbolic
Politics to Material Changes

Compared to the depletion issue, the water pollution issue (and more specifically the oil pollution issue) has moved more rapidly across our policy continuum. It had only a brief material non-decision phase from the end of World War II until 1947, when the Water Pollution Control Act was considered and passed in the Senate. This was followed by a period from 1947-1966, when the issue existed largely in the realm of symbolic politics. Water pollution legislation was passed but with little enforcement impact. Given the length of this period, it is not

surprising that this issue was selected as a good example of symbolic politics. The final period from 1967-1971 represents a shift from a symbolic politics phase to another material phase.

As these changes have occurred, the industry's strategy for dealing with the issue has also changed. And the corresponding symbolic stage for each of the issues reveals similar industry strategies. Differences lie in the level of success of these strategies for the industry. In turn, the level of success reflects the factors we discussed in the three previous chapters. An analysis of the progression through these periods will demonstrate the long-term failure of the symbolic politics strategy.

Non-Decisions: 1945-1946

This brief phase in some ways corresponds to the non-decision phase on the depletion issue. In both cases proposed legislation never went beyond committee consideration. But the causes are different as is the duration of the phase. Water pollution was not a major issue with which the Congress had to deal in the 1945-1946 period. Other post-war problems received the attention. While four water pollution control bills were introduced in 1945 in the House with companion bills in the Senate, none were reported out of committee during the 85th Congress.[32] In 1946 one minor control bill was introduced by Representative Mansfield (Dem., Tex.) and was reported out by Rivers and Harbors Subcommittee. No floor action was taken on the measure.

There are no accounts of industry opposition, attempts to stack committees, or delays of legislation. Water pollution was not a salient enough issue to receive serious consideration.

Entry into the Symbolic
State: 1947-1948

The passage of S. 418 by the Senate in 1947 and by the House in 1948 (in an amended form) marks the beginning of the symbolic politics period for water pollution legislation. This did not result from a long fight by industry groups to prevent any legislation, as with the depletion issue. Instead these groups supported S. 418. Conservation groups were also in favor of the legislation. Its sponsors, Senator Alben Barkley and Senator Robert Taft, insured strong Senate support for the measure. It was basically a public works measure authorizing over $20 million a year in the form of grants and loans for the construction of facilities (including $800,000 a year for five years for buildings and facilities in Cincinnati).[33] The enforcement provisions affected only interstate waters and were cumbersome to administer. Why should industry groups waste time

opposing legislation that seemingly would have little effect on them and had such strong backing? Karl Mundt, whom we noted earlier opposed the House version of S. 418, was quoted by *Congressional Quarterly* as saying:

I believe that this legislation will work to stop new attempts to write effective legislation, that it will protect present pollution practices, and that it will buy polluters additional time to practice their pagan program without being subjected to a workable formula for eliminating unjustifiable pollution.[34]

It would be hard to find a better descriptive example of symbolic politics in action. M. Kent Jennings, in his study of water pollution legislation, agrees that the 1948 act provided only "limited enforcement," but feels that the major impact of the legislation was in establishing "a foundation upon which subsequent legislation could be constructed."[35] These two opinions seem opposed but are, in fact, both correct. New legislation was written using the 1948 act as a base, but it did not come until 1956 (the 1948 act was renewed in 1953), and the changes in the enforcement provisions had only minimal impact. The new enforcement provision allowed the Surgeon General to call a conference when he had reason to believe interstate waters were being polluted and to recommend action after the conference. The polluting source was then given six months to take action. If no effort was made to meet the Surgeon General's recommendation, the Secretary of Health, Education, and Welfare could set a public hearing in the state where the pollution originated and require action after the hearing. Again, if no action were taken in six months, the Secretary could, if either state involved requested, ask that the United States Attorney General sue the polluter.

The 1961 act also marginally extended the enforcement provisions. It extended coverage to navigable waterways, thus including intra- as well as interstate waters,[36] and made it the discretion of the Secretary of HEW to call a conference instead of the Surgeon General.

While industry groups were active on these measures, Jennings finds that the major disputes in the passage of legislation during this time ran along party lines and involved the fiscal aspects of the program and states rights issues. Votes both in the House Committee and on the floor were highly partisan.[37] As the following tables show, opposition came from Republicans and some rural southern Democrats. Industry groups did object to broader enforcement in these proposals, and as noted in an earlier chapter, they worked primarily through Kerr and the Senate Subcommittee. These efforts were mainly geared to protecting states rights and limiting intervention on behalf of the federal government—goals with which Kerr was sympathetic. But during this time they were not geared to stopping legislation (in fact, they reacted favorably to some of the pork barrel provisions), only to keeping its enforcement impact limited. Naturally, their statements during public hearings regularly warned of disaster if the legislation were enacted. Charges of it undermining federalism were made regularly.

Table 5-1

Percentage of Votes in Support of Water Pollution Control on Related Roll Calls in the House of Representatives, by Party of Representative*

Roll Calls	Democrats	Republicans
1956 recommittal	88	21
1956 final passage	97	85
1957 appropriation deletion	84	23
1959 recommittal	89	9
1959 final passage	89	19
1960 veto override	90	10
1961 recommittal	92	14
1961 final passage	91	47

Source: M. Kent Jennings, "Legislative Politics and Water Pollution Control, 1956-1961," in Frederic N. Cleaveland (ed.), *Congress and Urban Problems* (Washington, D.C.: The Brookings Institution, 1969), p. 102. (©1969 by the Brookings Institution, Washington, D.C.)

*All percentages reflect the pro votes on water pollution control regardless of the specific direction of the original vote.

Table 5-2

Defection by Democratic Congressmen on Water Pollution Control Roll Call Votes

Roll Calls	Total Number of Defectors	Number Who Were Southerners
1956 recommittal	25	23
1956 final passage	5	5
1957 appropriation deletion	35	32
1959 recommittal	29	26
1959 final passage	28	24
1960 veto override	27	26
1961 recommittal	21	20
1961 final passage	22	21

Source: M. Kent Jennings, "Legislative Politics and Water Pollution Control, 1956-1961," in Frederic N. Cleaveland (ed.), *Congress and Urban Problems* (Washington, D.C.: The Brookings Institution, 1969), p. 104. (©1969 by the Brookings Institution, Washington, D.C.)

Enforcement evidence substantiates the case that industry groups had little to fear from this early legislation. In 1965 when Quigley appeared before the Muskie Subcommittee, he submitted information on enforcement proceedings since the beginning of the federal water pollution control program. Between 1948 and January 1965 there had been a total of 34 federal actions—21 federally initiated and 13 initiated by governors. Only four had reached the

hearings stage, and none had reached the court action stage.[38] Most cases remained at some exploratory study level.[39]

To get an idea of the time lags involved in enforcement, let us examine one of the cases that reached the hearings stage and where full compliance had been met. It happens, moreover, to involve oil well operations. On June 9, 1954 the Surgeon General, acting under the 1948 law, ordered correction of a situation in which discharges from oil wells endangered a drainage system in Louisiana. A public hearing was later held on January 16, 1957, after which the companies were required to cease the discharge of the brine runoff within 90 days of receipt of the hearing board recommendation. Full compliance was finally obtained in 1960.[40]

The symbolic politics approach appeared to be successful. However, by 1965 a realization began to grow among industry groups in general and the oil industry in particular that the symbolic politics strategy had some built-in dangers. The opposition groups were growing both in numbers and strength. Muskie now controlled the Senate action. No sooner was one bill passed than another one began. Reinforcement, not quiescence was the order of the day. And the 1965 bill proposed to make the next incremental step in the federal water pollution program. It would give the Secretary of HEW some authority to set water quality standards.[41] However, symbolic politics was not immediately rejected as a possible strategy. The public hearings on S.4 in both the Senate and House found industry groups making the same arguments they had made ten years earlier. But this time, in the shadow of civil rights legislation, there is a more serious attempt to emphasize the states rights aspects. This was especially true in the House hearings, where the conservative coalition remained strong in the Committee despite the 1964 elections. Rep. Cramer kept control of the Republican members and worked to get southern Democratic defections. His guiding of one industry witness is classic. The witness from the Manufacturing Chemists Association had just presented the industry position opposing Section 5 of S.4, which granted increased powers to the Secretary. Cramer, obviously feeling that the point should be reiterated, responded:

To me, this is the crux of the problem of the Federal Government trying to fix standards. It is amazing to me the lack of concern over the assumption of the terrifically broad, general power never before assumed by the Federal Government, under the guise of setting standards for water.[42]

Several comments later he led the witness with the following rhetorical question about the Secretary's power:

It seems to me it would be an awfully open blank check for him to do just exactly what he wants to do regardless of what the states may think about it in the setting of the standards.
Do you agree with that point of view?[43]

While earlier legislation had evoked the same sorts of responses from industry, the potential impact of that legislation was small. Federal intervention was limited, and no clear pollution standards existed except those set at the state and local level. The 1965 legislation seemed to be an incremental change that had hit a threshold point. For the first time there existed the potential for material consequences. The 1965 legislation marked a change in the focus of enforcement. And if the nature of industry response had not changed, its intensity had. Sundquist reports that the House Committee was faced with "industry threats to kill the entire bill, as in 1964, if federal water standards were retained in it. . . ."[44]

Moreover, the nature of the 89th Congress no doubt heightened industry concern. With 295 Democrats in the House, there was no chance of even a close floor vote on any water pollution control legislation, as there had been in the 1956-1961 period. However, the conservative coalition still maintained a potential 18-17 majority on the Public Works Committee. The bill, as reported out of the House Committee, left water quality standards as a state responsibility with federal power available only if states did not file intentions of setting standards by the end of fiscal 1967.

Although industry groups had previously obtained compromise solutions, none was of the magnitude of this one. It was meant to stop a threshold from being crossed. And while it still represents the willingness of industry groups to play the symbolic politics game, it also indicates the inevitable failure of that strategy. When symbolic reassurances reinforce rather than quiet the opposition, the continued use of them is likely to have material consequences. The issue area then takes on more of a mix of symbolic and material politics. While symbolic reassurances are continually offered in the form of new legislation, the impact of these symbolic increments becomes of real importance. The issue, in effect, moves from a symbolic to a material policy stage.

In water pollution legislation, this meant that industry strategy had to change. At some point during the 1965 bill it became clear to certain industry groups that the continued reliance on symbolic reassurances had to cease. Instead, the attitude became one of crippling or defeating new legislation detrimental to the industry, at all costs. In 1965, when the House and Senate conferees met, this change in strategy meant a new level of adamancy on the part of industry supporters. They would not settle for a bill that would take primary responsibility for setting water quality standards away from the states. Some conservation groups were willing to accept a compromise making the federal role a secondary, coercive one.[45] Muskie held out for a federal prerogative to impose standards if the states failed to do so or did so inadequately in the mind of the Secretary of HEW. The conference took better than four months. Hard political bargaining had been added to symbolic reassurances, even if it had not replaced them.

The law had passed, but no effective change occurred in the federal enforcement program. All states set standards and the federal government's ability to intervene was effectively undermined. In 1971 Elmer B. Staats, head of the General Accounting Office, testified before the House Public Works Committee. He claimed that the federal government was nearly powerless to act on water pollution except in cases of interstate pollution, and then only when the pollution became a problem. Even then enforcement procedures under the 1965 act were so cumbersome that a minimum of 58 weeks was required between the time EPA held a conference to decide on corrective action and the date when the agency can request the Attorney General to take legal action.[46]

The oil industry was one of the first industry groups to reject a pure symbolic politics approach. In fact, it is difficult to get oil lobbyists to admit that they ever fully accepted it on pollution matters. One argued that perhaps from hindsight you would be better off taking a small loss early if it could prevent a larger one later on. But he claimed this did not work in the pollution area. He elaborated, "Even if it drains the people who got the first bill through, there will be others pushing new legislation." He felt there was no longer anyone at API who used this strategy on pollution issues. There had been one person who believed in this method of operation, "but he retired and not because he was old enough either." An oil lobbyist for one of the major companies offered a similar opinion:

I take them as they come and try not to be a Monday morning quarterback. If you stop it at one point it's always possible something else will come up and take its place. I'm not one who believes a compromise bill today will prevent opposition tomorrow—especially on pollution. You can't legislate its end, and they'll keep coming back after you.

An EPA official saw the oil industry as one of the groups that continually fought all out in opposition to water pollution bills.

The 1966-1970 period clearly demonstrates this attitude on the part of the oil industry. It was during this time that water pollution bills contained specific sections on oil pollution. The industry used every possible opening to delay, defeat, or at least undercut the impact of the proposed legislation. In 1966 it took the form of the Wright amendment in conference;[47] in 1968 the opening came when Muskie was campaigning on the day the Senate adjourned; and in 1969-1970 a coalition of House conferees delayed conference agreement for nearly five months.

Each time a bill was defeated or undermined the stakes were increased as the committees reported out stronger legislation. In this sense the industry strategy seems a little shortsighted. The provisions of the 1966 legislation even before the Wright amendment were far less severe on the industry than those of the 1969-1970 bill. Even the 1966 Senate version provided for small maximum criminal penalty of $2,500 and one year in jail for oil discharges from vessels and

a $10,000 civil fine for discharges from vessels and shore installations. The 1970 law provides for "absolute liability" regardless of negligence for oil spills from vessels and onshore or offshore installations, with liability limits of $14 million or $100 per ton (whichever lower) for vessels and $8 million for offshore and onshore installations.

But is this really a shortsighted strategy? The industry hoped that the pollution issue would be displaced by some other issue. And even if it were not displaced, accepting legislative increments would not prevent new efforts to go after the industry. Following the 1965 bill, the industry witnessed the failure of the symbolic politics strategy. The testimony of David L. Gallagher of the National Association of Manufacturers before the Muskie Subcommittee in April 1966 is illustrative:

Keeping in mind that it has been only 7 months since the Congress, after several years of deliberation, passed the Water Quality Act of 1956, intended to 'enhance the quality and value of our water resources,' some of the proposals in S.2987 seem reminiscent of the gardener who kept pulling the plant out of the ground to see if the roots were growing.[48]

As Sundquist points out, the Muskie staff had begun work on the 1966 bill before action on the 1965 bill was completed.[49]

Part of the shortsightedness of the strategy is more apparent than real. The issue, rather than being displaced, became more salient due to the major oil spills in the late 1960s and the media coverage they received. Prior to April, 1967 news reports of oil spills from vessels and onshore and offshore installations were rare. True, the spills were not as frequent as in the 1967-1970 period. But they did occur. After the 1970 fire and spills from 12 Chevron wells off the Louisiana coast, Lt. Governor Aycock of Louisiana claimed:

We have had pollution from oil in south Louisiana for 20 years, but now we are suddenly beating our breasts about it. We are, in effect, killing the goose that lays the golden egg.[50]

After the unsuccessful cleanup of the Torrey Canyon spill in early 1967 off the coasts of Britain and France, press attention to oil pollution skyrocketed. After Torrey Canyon even minor spills were reported. *The New York Times*, for example, devoted a series of stories to a small oil slick off the New Jersey coast just after the Torrey Canyon incident.[51] Other larger spills received more attention. And late in April *The Times* released the story on the inadequacy of the 1966 amendments to the Oil Pollution Act.[52]

It appeared that the oil pollution issue might subside in late 1968; the industry had been successful in preventing House-Senate agreement on oil pollution legislation.[53] But in January 1969 the blowout on the Union Oil drilling operation in Santa Barbara Channel gave the issue a new boost. Once again, even small spills received press attention. And, as before, there was a

gradual drop in this coverage toward the end of 1969. Then, in early 1970, while the conference committee was meeting to iron out differences in the House and Senate versions of the 1969 water pollution legislation, a new series of major spills occurred. The most important of these, already discussed, was a tanker spill which covered 25 to 100 square miles of Tampa Bay in mid-February. But an even larger spill was caused by the wreckage of the Imperial Oil tanker *Arrow* off Nova Scotia. The *Arrow*, with only a 3.8 million gallon capacity and only a third as large as some of the super tankers, was a blunt reminder of the need for higher liability limits.[54] In addition, the Chevron fire also took place in February. It was no surprise, then, that the conference adopted the absolute liability provision of the Senate bill.

The strategy cannot solely be judged by the industry's effectiveness with Congress. Once legislation with potential material impact was passed, the strategy placed increased emphasis on the enforcement agencies. When legislation had only symbolic potential in the first place, the industry groups involved themselves with enforcement on the same *ad hoc* basis on which the enforcement program existed. The 1970 act, with its absolute liability provision, threatened across the board enforcement. The law covered "harmful quantities" of oil and required that the President determine what constituted a harmful quantity as soon as possible. The definition of a harmful quantity thus became the point of concentration of the industry's effort. All indications are that the industry was successful in delaying final definition of the term. In June 1970 the White House met the other requirements of the act but left the act inoperative when it failed to define "harmful quantities."[55] In late July, Secretary Hickel proposed that an oil discharge big enough to cause a visible slick be considered harmful and allowed 30 days for comment on this decision.[56] Senator Muskie responded in early August that the Interior criteria were so vague that they would circumvent enforcement and that another new regulation obliging the spiller to notify authorities would reduce the government's ability to seek civil and criminal action against the spiller. This followed a decision by Interior to rewrite the regulation after the industry protested the definition.[57] On November 31 *The Times* reported that the oil spill definition would once again be delayed. President Nixon, the article claimed, had planned to relax the regulations. Fred Russell, the acting Secretary of the Interior during the transition from Hickel to Rogers Morton, was to have made the announcement. But the decision was delayed pending the confirmation of William Ruckelshaus as head of the new Environmental Protection Agency.[58]

At hearings before the Muskie Subcommittee in February 1971, EPA reported on the status of regulations that it was promulgating under the Water Quality Improvement Act of 1970. At that time, the agency claimed that Section 11(b)(3) of the act on the subject defining harmful discharges was under review in the General Counsel's Office of EPA. It was not until November 25, 1971, that a final definition was placed in the Federal Register. During the

intervening period, industry officials were working for a lenient definition. But EPA, they complained, was being unrealistic. One lobbyist said it was impossible to define a "harmful" quantity. He saw the industry as getting nowhere with EPA on this point:

They want it to be any discharge which results in a sheen. Well, heck, a capful of oil could cause a sheen the size of this deck. Companies report everything to the Coast Guard. The Coast Guard asks how big it is and the company says it's the size of a floor rug. They laugh and say forget it.

The final definition included discharges which:

a. Violate applicable water quality standards, or
b. Cause a film or sheen upon or discoloration of the surface of the water or adjoining shorelines or cause a sludge or emulsion to be deposited beneath the surface of the water or upon adjoining shorelines.[59]

Thus, one year and eight months passed between the signing of the 1970 act by President Nixon and the finalizing of the regulations of the act. During the first year of this period EPA reported that there were 40 major oil spills.[60] While David Dominick was able to claim "notification of oil spills has increased dramatically from 1969 to 1970 by virtue of the passage of the Water Quality Improvement Act of 1970,"[61] he admitted that the Coast Guard regulations on reporting oil spills were not published until November 21, 1970, so that enforcement of the notification provision would have been impossible before them.[62]

Unfortunately for the industry, the delay in establishing these regulations did not stop FWQA (and later EPA) from taking enforcement action in regard to oil spills. Instead of using the 1970 act, federal officials relied on other water pollution laws—specifically, the Rivers and Harbors Act of 1899 (often referred to as the 1899 Refuse Act) and the Outer Continental Shelf Lands Act of 1953 (as amended). The oil industry offered opposition to the use of each. The 1953 act gave the Secretary of Interior certain controls over leases of offshore lands. In August 1969, in response to the Santa Barbara spills, Secretary Hickel required that offshore wells be equipped with either storm chokes or other safety devices that could close off a well in case of accident. When the Chevron fire occurred in February 1970 violations of this requirement were discovered. After some public prodding from Senator Muskie,[63] Secretary Hickel requested the Justice Department ask for indictments under the act. In May, a federal grand jury indicted Chevron on 100 counts (the act called for violations on a daily basis).[64]

The industry responded to this not only by pleading "not guilty" to the charges but by activating Louisiana residents and political leaders. While Lieutenant Governor Aycock objected to the federal action, which included a decision to stop leasing properties in the Gulf as well as the court action,

Governor McKeithen volunteered to defend Chevron.[65] In addition, a bumper sticker campaign began in the state. The stickers read: "Oil Feeds My Family."

Again, however, the industry found itself on the losing side. On August 26, Chevron was found guilty on each of 500 counts and was required to pay a fine of $2,000 per count or a total of $1 million. The remaining 400 counts were dropped. Also a check of other offshore operations found violations of this same regulation by Humble, Union, Continental, and Shell wells. When the cases came to court on December 2, only Shell pleaded "not guilty." The three others claimed no contest and paid fines of $300,000, $24,000, and $242,000 respectively.[66] On December 22, the Justice Department arrived with early Christmas presents for Gulf, Mobil, Tenneco, and Kerr-McGee. It sued these four companies for 226 violations of the 1953 act.

When this act was not applicable, in cases of oil spills from vessels and onshore installations, EPA relied on the 1899 Refuse Act. The act provides for fines of from $500 to $2,500, or imprisonment from 30 days to one year, or both for depositing refuse in navigable waters unless under permit by the Secretary of the Army (on the judgment of the Chief of Engineers).[67] (It also provided that one-half the fine be given to "the person or persons giving information which shall lead to conviction.")[68] During 1970 the Refuse Act was applied 365 times in cases of discharge of oil on navigable waters. Most of these actions were taken in non-oil producing regions by EPA and the Corps of Engineers (see Table 5-3).[69] Of these, 65 resulted in fines of $500 to $2,500. On most, action was pending. In no case was the imprisonment option used. Most were small spills and were not on the EPA list of major oil spills. And although the fines were similarly small, the oil industry was upset with the invocation of the act. One industry spokesman expressed vehement feelings that the 1899 act was being misused. He claimed that it was being used to sidestep the decision

Table 5-3
Refuse Act of 1899 Passes in 1970

Region	Total	Oil	Non-oil
South Central	39	32	3
Middle Atlantic	23	21	2
Southeast	4	3	1
Missouri	5	0	5
Southwest	1	1	0
Ohio Basin	63	30	20
Northwest	39	16	18
Northeast	505	257	248
Great Lakes	24	5	19
	703	365	316

process of the 1965 bill, which gave the states authority in setting standards "only then to have EPA come along with the Refuse Act and say they're not any good."

What really upset him was in all likelihood not just that EPA or the Corps could step in, but that the 1899 act enabled the government to get an immediate enforcement in the form of a court injunction while the 1965 act had the lengthy enforcement pattern.

All this does not indicate that the oil industry was necessarily shortsighted. Rather I think it indicates that the industry was left with no other short-run strategy to pursue than to try to win at each decision making point. When success could no longer be found in efforts directed at the legislation itself, the industry merely moved its efforts to the administrative agencies.

In this way its strategy was similar to the one used prior to 1969 with regard to depletion. As with depletion, the foreseen consequences were highly material and a strategy of all-out opposition was used. The vital difference, however, is that all-out opposition to depletion resulted in a non-decision strategy. But with water pollution the same all-out effort was the tool of last resort. At best it could only serve to delay enforcement of oil pollution legislation. The industry could not stop legislation from being reported out of committee as with depletion, nor could it afford to ignore the consequences of the legislation once it was finally signed into law, as in the symbolic stage of pollution legislation or after passage of the Tax Reform Act of 1969. In those cases, the consequences of the legislation were sufficiently small or could be avoided without major bargaining efforts with federal administrative agencies.

Naturally, these stages—material and symbolic—are not totally separable, neat phases. While it has been useful to talk of them as if they were clearcut, there is a mix, albeit a changing one. Thus, in the post-1965 period the water pollution issue continues to have symbolic political components. While fighting all-out against technical enforcement provisions, the oil industry was quite willing to have the impression fostered that the government was fighting pollution. Industry officials are ever ready to correct anyone who implies that they are opposed to the government's efforts in the pollution area. They always emphasize their concern with the water pollution problem. But in the post-1965 period the mix is a very different one. No longer is the industry willing to allow new legislation to pass without a fight. While mass public reaction may remain unchanged—and still respond to the symbolic quality of legislation—industry activity indicates that policy is changing.

The return to the all-out strategy indicates that: (1) the consequences of legislation include a clear material component and (2) the industry position in the issue area has deteriorated. Again the main resource the oil industry has had working in its favor since 1966 on water pollution legislation is the serial nature of the process. Delay or defeat for legislation may come by winning at one point in the process. Attention progressed from a focus on the House Public Works

Committee to the Conference Committee to the administrative agencies (and the President).

Issue Movement: A Generalized Form

We can better understand what occurs as an issue moves from one policy level to another through the use of Chart 5-1. What it shows is the effect of incremental policy change on the strategy, tactics, and success of the oil industry as generalized from examination of our two issue areas. We have distinguished three policy phases: a non-decision phase, a symbolic phase, and a material phase. These are consistent with Edelman's description of "symbolic politics." The non-decision phase is merely a refinement of Edelman's theory. As an issue moves from one stage to the next, in a left to right fashion on the chart, the industry (or private interest group) position is deteriorating. At each phase the industry's effort was designed to keep the issue at or move it back to the non-decision phase. The depletion issue has moved from the non-decision to the symbolic stage during the period of study, while the water pollution issue has gone from the non-decision to the symbolic to the material stage. As policy making for each issue moves along the continuum, industry strategy and tactics change as do the foci of decision making.

In Chapter 1, I claimed that I had selected the issue areas because they appeared to fit Edelman's criteria for symbolic and material politics. And during most of the time period we considered, the issues were congruent with his distinctions. Only in the late 1960s and early 1970s did water pollution legislation begin to supplement symbolic politics and with some material impact. Similarly the depletion issue remained largely an area of non-decisions until 1969, and only then did it take on a symbolic character. The material label applied in Chapter 1 is the correct one, since the consequences of non-decision politics are in fact material and do not provide symbolic reassurances. This is, however, a different type of material politics than exists in the third stage.[70]

The important point is that issues do not necessarily remain at one particular stage but may, in fact, move. In the cases we have studied movement has been in one direction. But incremental changes can be in the opposite direction, just as incremental budgeting can mean cuts as well as increases.

Factors that Affect the Direction

The three previous chapters have dealt with some of the factors that affected the industry's ability to influence both the direction and speed of policy change. To the degree that the industry has strong constituency ties at the relevant committee and subcommittee level, it is able to resist movement toward material

Chart 5-1
Interest Group Activity

	Incremental Policy Changes		Material Policy Changes
	Non-Decision	Symbolic Policy Changes	
Tactics	Activate constituency ties to committee members, especially chairman and staff.	Speak out in public against legislation. Bargain in private.	Concentrate on each episode in process. If resources relatively unlimited, try to stop legislation at each. Otherwise, concentrate where chances best.
Strategy of Private Interest Group	Defeat legislation early.	Allow legislation to pass. Keep impact minimal through efforts in committees.	Try to defeat or delay legislation or to delay its impact.
Focus of Activity	Relevant Congressional committees.	Still the committees with some attention to admin. agencies after legislation passes.	Conference Committee and administrative agencies (after passage).
Potential Consequences for the Industry	None	Few, if any.	Major material impact.
Potential Consequences for Opposition (if one exists)	None	Potential for either quiescence or reinforcement from symbolic reassurances. No objective policy impact.	Major objective changes.

policy change. This was the most important factor in keeping the depletion issue at a non-decision level. It can keep the issue at a sufficiently low visibility level that it is unnecessary to offer any symbolic reassurances. Naturally, changes in the level of such ties can have policy impacts. The clearest example we have of this is the change from Kerr to Muskie as chairman of the Senate Subcommittee handling pollution issues. Not only did this mean the industry redirected its effort to the House Committee but also it resulted in the Senate's producing legislation that had potential material impact.

Rules, procedures, and processes we found generally to work in favor of non-decisional policy and thus in favor of the industry. Again, changes in these had impact for policy. The more open recruitment of Ways and Means Democrats after 1961 is a leading case in point.

Finally, examination of the organization and abilities of opposition groups demonstrated a clear relationship to these policy movements. In the case of this factor some mutual reinforcement did occur. Symbolic reassurances did not quiet those involved. In the water pollution area, incremental policy changes are associated with increased group involvement, which in turn resulted in additional policy changes. Nevertheless, factors outside the isolated political environment of each issue led to increased attention of opposition groups. Oil spills, air inversions, and general increased levels of pollution made the pollution issue one around which support could be rallied, and increased taxes associated with an unpopular war had a similar effect for tax reform.

In addition, elections, which we have not directly considered here, have had significant impact on policy change. Sundquist documents this for a series of domestic policy areas. He argues that elections are a major force in policy change despite the non-policy-oriented American political parties.[71] Our clearest example of this impact is provided by the change in Treasury's attitude toward tax reform after the Nixon Administration took office. Rather than leading the fight for reform as during the last years of the Johnson Administration, Treasury dragged its feet on the reform proposals.

On the issues of water pollution control and the depletion allowance, these factors during the last two decades have tended to change in directions that weaken the oil industry's position. As long as these changes continue to move in this direction, we can expect corresponding policy changes to occur in each area. But there is still reason to believe that the industry position is a salvageable one, especially in regard to the depletion allowance. The industry's ties to the Ways and Means Committee and the Finance Committee are as strong, if not stronger, than they were in 1969. Hale Boggs left the Committee to become majority leader, but was replaced by Joe Waggonner of Louisiana, who should prove far less compromising on the depletion issue.[72] The replacement of John Watts with Joseph Karth of Minnesota should provide some counterbalance. But if the Committee's split on the issue is as close as the 13-12, our scalogram suggests that the Boggs-Waggonner change could be decisive.

The Finance Committee poses a different situation. Strong anti-depletion spokesmen Albert Gore and John Williams have left the Senate. Robert Griffin of Michigan replaced Williams and Gaylord Nelson of Wisconsin replaced Gore. Neither successor can be expected to be as vehement on the issue as his predecessor. Two respondents claimed that Nelson was a "social buddy" of Russell Long's, and that unless the issue became "very hot" Nelson was expected to be safe.[73] Moreover, a continuation of the Nixon Administration after the 1972 election would again provide the industry with strong ties at Treasury. Testifying before the Ways and Means Committee on May 1, 1972, Edwin Cohen indicated that Treasury was opposed to broad tax reform.[74]

There is one added complication. No one has a good estimate as to the level to which the depletion allowance would have to be cut before major segments of the industry would feel the impact of the change. The former Treasury official whom we quoted as claiming that oil companies with foreign holdings do not now use all the foreign tax credit available to them feels that until depletion is cut to 15 percent no substantial impact will occur. It may take several more tax reform bills before that threshold is reached.

In the oil pollution area the struggle has shifted. The issue is no longer over the liability for oil spills and who will perform cleanup operations. The companies have lost the struggle here. Environmental groups are now trying to prevent the federal government's leasing of additional offshore property, while the oil companies have tried to demonstrate their willingness to cooperate in cleanup efforts. In October 1971 the American Petroleum Institute held a conference to show how the industry was responding to public concern over oil pollution. One report of the conference claimed "when participants sat down for discussions in New Orleans, it was hard at times to tell the conservationists from the oil men."[75] In November 1971 Union, Gulf, Mobil, and Texaco agreed to pay $4.5 million to property owners whose beaches were polluted by oil from the 1969 Santa Barbara blowout.[76]

But environmental groups have not been satisfied by these responses and have interfered with federal leasing policy. In Santa Barbara a group called Get Oil Out has been working for the termination of oil leases in the channel.[77] The Natural Resources Defense Council, Friends of the Earth, and the Sierra Club sued to block the Interior Department's leasing of 78 offshore tracts in the Gulf of Mexico in December, 1971. The sale was delayed when U.S. District Judge Charles R. Richey issued a temporary injunction forcing the Interior Department to delay the sale.[78] During that same month East coast Senators and Representatives met with Secretary Morton to express concern over plans to lease areas on the Atlantic outer continental shelf for oil exploration.[79] This struggle has continued and is still far from resolution. In addition to these, the long running dispute over the Alaskan pipeline has been very costly to the industry.

We cannot yet tell who will win these struggles. But the stakes, while containing some symbolic overtones, are clearly material in their potential consequence.

A New Strategy for Oil

In the post-1969 period the oil industry has geared itself to a new strategic response to the tax and pollution issues as well as other issues. In place of specific tactics applied to each separate issue, the industry's new approach is a general one meant to protect it on a broad range of issues. The name most commonly used to describe the thrust of this approach is "the energy crisis." It is not a new line of argument. Throughout the period under study industry spokesmen regularly refer to the need for a large supply of inexpensive domestic petroleum. This line of reasoning has recently taken on new elaborations. Not only is the public warned about the need for oil reserves, but it is suddenly told of the fact that reserves are being used at a more rapid rate than current operations can replace them and that an energy shortage is at hand.

Without trying to evaluate the validity of the industry's claim, we merely note that the industry relies heavily on it in attempting to maintain both its tax position and its offshore drilling operations.

An additional change has occurred with the industry's adoption of this stance. Instead of merely making claims of increasing fuel needs in testimony before Congressional committees, the industry efforts have taken on a public relations aspect. In early 1971 an API official claimed that the industry was finally seeing the need to make its case publicly. Since that time API has launched an advertising campaign. The campaign includes extensive use of television to inform the public about "energy gap U.S.A."

Every industry spokesman with whom I talked thought that the industry's position was stronger on both the depletion and pollution issues than it had been in 1969 because of the threat of an "energy crisis." An IPAA lobbyist, in arguing against the strategy of accepting small changes to prevent bigger ones, noted that the change in the depletion allowance offered no "evidence of its pacifying anyone who was against us." But he claimed, "If there was an attempt to cut depletion now instead of in '69, we'd be in much better position" because of the energy crisis.

Since I conducted these interviews, the threat of an energy crisis has become much more critical. The gasoline shortages in the summer of 1973, compounded by the Arab-Israeli War and subsequent oil embargo, have resulted in conspicuous activity to deal with the energy crisis. Be it a real or imaginary crisis, it has affected both governmental activity and the daily lives of individuals. Speed limits have been cut, thermostats turned down, and gasoline stations closed. We are presented with a new energy office, regular Presidential messages on the crisis, and daily estimates of fuel oil and gasoline. But the most important consequences have been the tremendous increase in the price of oil and the increased profits to the oil industry. A staff report of the Senate Permanent Investigations Subcommittee has been cited as showing that for even major oil companies, total sales volume increased 6 percent for the first three quarters of

1973 compared with the same period in 1972, while total revenues increased 22 percent and total net earnings 46 percent. Using the 1971-1973 period to adjust for what the industry claimed had been a poor financial year in 1972, earnings still increased on an average of over 20 percent a year over the two years.[80]

In more general terms, the new strategy is an attempt at substitution of conflicts.[81] The oil industry is trying to make the energy crisis issue dominant over both the issues of tax reform and water pollution. This strategy, as with the symbolic politics strategy, recognizes that socialization of conflict exists. And like that phase, it proposes to deal with the problem by splitting the other side (through the introduction of new issues) and by offering the symbolic reassurances of sufficient energy supplies.

While it may be too soon to judge the effectiveness of this strategy, especially given the additional complexity of the Arab oil embargo, some basic boundaries can still be drawn. First, the strategy attempts to affect the entire agenda. To the degree that it is successful, tax and pollution matters should no longer be viewed in their narrow issue contexts but rather in the context of energy needs. And as with non-decision politics, the effort is designed to bring victory for the industry at an early stage in the process through definition of what the issue really is. If the strategy is fully successful, we can expect a return to a non-decision phase. The opposition will be split and the issue one on which decision makers can easily support the industry position. Just as tax reform gained dominance as an issue in 1969 over the need to encourage mineral exploration, another transaction can occur.

Second, as incremental changes proceed on each issue, the substitution of the energy crisis focus may become easier. One former Treasury official saw this as a likely possibility. He felt that Congressmen who would willingly vote to cut depletion to 20 percent for the sake of tax reform would become more responsive to arguments of fuel shortages and thus more amenable to higher prices for the consumer as additional inroads were attempted.

It is difficult to place this strategy as a reaction to an issue's being symbolic or material in potential impact. It can be used against any issue that threatens the industry. Yet the energy crisis clearly has symbolic overtones, and ones with which the industry thinks it can be successful. The fact that the events of 1973 have made the crisis more than symbolic merely makes the symbolic threat more credible. In any case, the strategy demonstrates the oil industry's recognition of the difficulty it has had in keeping any of its operations outside the public arena in recent years. Further, incorporated in the strategy is the realization that, due to the growth of organized oppositions, winning in the public arena would be impossible unless major segments of the public were either neutralized or won over to the industry's side. It does not matter whether the energy crisis was explicitly created by the industry to assist its lobbying efforts or is in fact the result of other causes. In the short run, it will be used to protect the oil industry's position on the two issues investigated here, as well as on a series of

other issues. The employment of this strategy is the final admission of the ever-weakening industry position on these issues. The industry can no longer play non-decision politics. And even limited symbolic rewards are unlikely to bring about mass public quiescence. Rather, the industry has been forced to accept the fullest form of conflict socialization. While this may succeed, that success, as will be discussed in the final chapter, is far less predictable.

Summary and Conclusions

From the preceding analysis, it is clear that policy type contributes much to our understanding of the oil industry's activity and success in each of the two issue areas above and beyond that offered by the variables we examined earlier. The major contribution of policy types is in providing a source of explanation for changes in the industry's strategy. Strategies reflect the movement of an issue along the policy continuum we have adopted. Thus, the industry's use of symbolic reassurances follows from the fact that a particular issue has moved to a point of public visibility where decision makers can no longer respond favorably to non-decision tactics. New tactics must be used to deter further movements of the issue. A change from offering symbolic reassurance to fighting "all-out" to defeat legislation similarly reflects the failure of reassurances to stop the issue's movement and the potential for material impact to be achieved from additional movement of the issue.

Naturally, the relationship between policy and strategy is not a pure one. Especially during transitional phases, like the 1965-66 period with the water pollution issue, a mixture of strategies often exists. And since certain segments of a given issue may reside at different policy levels, some strategy mixes are likely to be present even when the total issue appears to fall neatly within one policy level. At best Chart 5-1 shows the tendency that exists as an issue moves across the policy continuum.

A second feature that protrudes from this analysis is that as the issue moves from the non-decision to the symbolic to the material stage, the industry places an increased amount of its effort on the Executive agencies. True, in the non-decision phase some efforts may be devoted to preventing administration proposals. But beyond this industry concentrates on the Congressional committees and subcommittees. Later, as legislation begins to pass the Congress, increased attention is given to the enforcing agencies. This has happened most clearly in the pollution issue, and we might expect a similar effort in the depletion areas, i.e., the industry will work with Treasury after legislation passes and not just before.[82]

The third benefit derived from the analysis of policy types is that the policy state at which an issue resides indicates the general success with which the defending interest is able to maintain its position. The more rapid movement of

the water pollution issue across the policy continuum, when compared with the movement of the depletion issue, demonstrates this point. Moreover, it leads us to look at factors like constituency strength; rules, procedures, and processes; and the nature of the opposition for explanations of the differential success.

Aside from these obvious advantages, the policy continuum supplies us with certain perspectives for the analysis of interest group behavior. First, group actions can be analyzed in a more systematic light. The strategies no longer appear as meaningless as some of the works on interest groups would have us believe.[83] Second, the framework allows for comparison of issues in a meaningful manner. Issues are no longer just apples and oranges, but instead can be viewed with the focus of a common denominator.

Finally, this focus stimulates additional questions for the study of interest group politics. Does the same policy ordering operate when an interest group is proposing rather than opposing new legislation? Will the incremental policy process be altered as more, better-skilled public groups become active? Is the focus of interest group attention continuing to move to the Executive agencies, even in areas, such as tax legislation, where Congressional influence has previously been predominant?

6   Where Do We Go From Here?

After digesting numerous pages about the oil industry's activity on water pollution and depletion allowance legislation, the reader may properly ask where the analysis of this activity has led us. Of course, there were conclusions drawn at the end of the research chapters. But beyond these, what specific conclusions about the present and future success of the oil industry's lobbying activities can be drawn from this research? And more importantly, what are the broader implications of this study for research on interest group behavior in the Congressional process, and what questions does it raise for future research. The purpose of this conclusion is to address itself to these three points: the future of oil industry behavior, the broader implications of this study, and questions for future research.

First, it is quite clear that the position of the oil industry in influencing legislation has weakened considerably in the post-World War II era, and especially in the late 1960s and early 1970s. While two of the factors analyzed here have remained fairly stable—constituency ties and rules, procedures, and processes—change in both the nature of opposition and the policy content of the issues have been seriously altered. In both cases, the changes have limited the strategic options available to the industry. The new public stance the industry has taken is not a sign of strength, but a symptom of weakness. If the industry were truly as powerful as earlier reputed, it would have no need to resort to a public appeal. Decisions affecting the industry's tax situation and environmental controls on drilling operations would instead be made at low visibility decision points. Now the industry must depend on favorable administrative decisions and on controlling the issue as it is presented to decision makers. To the degree that decision makers and the public are persuaded that an energy crisis does indeed exist, the industry should continue to have success on a variety of legislative matters. But this poses new problems that go beyond mere control of the issue at hand. By accepting the public stand of the energy crisis strategy, the oil industry invites public and governmental awareness, interest, and investigation into a broad range of the industry's operations. While in the short run it may enable the industry to avoid certain environmental constraints in drilling for oil and to protect and even add to tax provisions designed to stimulate exploration for oil, the consequences of the strategy may also mean further investigation of industry tax privileges (especially the depletion allowance), tighter governmental control of drilling, refining, marketing, profits, and a general increase in overseeing previously unregulated industry activities.

We should not think that these problems are in any way insurmountable for the industry. It still controls most of the information on its operations and the expertise for interpreting this information, and those are indeed valuable resources. Moreover, given the long history of the regulatory "life cycle," one may question whether increased government interference in industry operations will in the long run prove an undesirable consequence as far as the industry is concerned. Nevertheless, this is a marked change from the narrow political environment in which the industry operated through the 1950s and much of the 1960s, when the oil industry found that everyone was a "friend" on Capitol Hill.

The most important implications of this study, however, stretch far beyond conclusions about the future of oil industry lobbying success. They deal with the systematic understanding of interest group behavior. First, the findings set boundaries for understanding the activities and success of interest groups in the Congressional process. The conclusions of the present study should hold for other groups and issues. We have seen that the oil industry's actions and success have differed in systematic ways with the strength of constituency ties, the visibility of the decision making point, the ease with which opposing groups enter and follow the Congressional process to conclusion, and the policy components of the particular issue at a given point in time. The industry's actions have not been without rhyme or reason. They have varied in accordance with these and other factors. While no exact values can be established for each of these factors, some of their parameters have been identified and the nature of their impact spelled out. Thus, it was found, for example, that the existence and effectiveness of a mass opposition depends on the type of goods involved in a given issue, the complexity of the issue subject matter, and the visibility of the problem around which the issue is focused.

A second key implication is the association of policy variability with changes in the tactics, strategy, and success of interest group efforts. The findings here reinforce the notion that issues do not always have the same policy components, but tend to take on new policy mixes over time—and that dependent on the policy mix, an interest group's tactics, strategy, and success will vary. In fact, much of an interest group's efforts are likely to be devoted to keeping an issue at or moving an issue to a given policy phase. Thus, the oil industry's efforts were geared first to keep the depletion and water pollution issues at a largely non-decision policy phase. As these issues took on new policy components, the industry's efforts were geared to returning the issue to the non-decision status. However, as these issues took on symbolic and finally material policy components, the difficulty of returning to the non-decision phase increased.

The third implication of this study, directly related to the preceding one, involves the development of policy typologies. Whereas policy has been shown to be an important independent variable, there is a need for further development and refinement of policy typologies and continua. Edelman's symbolic-material conceptualization was certainly valuable for the current study. But despite the

desire expressed in Chapter 1 to use Edelman's dimensions (in part, because of their simplicity and allowance for mixes), it was still necessary to distinguish between two types of material politics in Chapter 5. Moreover, while it is useful to think of some policy stages as mixes, it would also be helpful to develop criteria that discriminate among various mixes. In this study, mixes were distinguished from each other largely on the basis that they were located at points closer to one end of the continuum than other mixes. In this sense the symbolic-material politics mix in the water pollution issue in 1965 could be said to have had less material component than the mix for the same issue in 1969. The development of greater specification here would clearly be an improvement.

One caveat should be made at this point. While the study has emphasized the importance of policy as an independent variable in understanding interest group behavior, this has not been done with the intent of diverting attention from other variables. If anything, the reverse is true. In explaining interest group behavior in the Congressional process, we cannot simply exclude from our focus the variables more traditionally used for this purpose. If anything, this study indicates that, despite their long usage, many of these variables have not been fully explored. (This is especially true in the evaluation of the impact of opposition interest groups, though even the effects of rules and procedures remains relatively untapped.)

We cannot settle for the broad level of explanation Truman provides in Chapters 11 and 12 of *The Governmental Process* as sufficient, although it provides an important starting point. True, knowing that a given interest group has access to a particular decision making point is important in understanding its success on a given issue. But knowing how that access was established, under what circumstances the access will change, what other groups have access there, and what alternatives to access at that particular decision point are available for the interest group enables us to understand interest group behavior and success in a more dynamic sense. It is to these questions among others that Chapters 2-4 were directed.

We must remember that the variables examined in those chapters were designed to lay the foundation for examining policy as an independent variable. Without that a concise analysis of policy impact on interest group behavior is futile. Rather than distracting our attention from process variables, the examination of policy as an independent variable requires that we sharpen our focus on them.

The fourth implication of this study is somewhat tangential to the specific research area at hand. That is: good research in many areas of political science requires the evaluation of public policy and policy changes. To perform the research in a meaningful fashion requires that the researcher himself become a policy expert. While I would not claim that I am now an expert on all facets of taxation of mineral industries or on water pollution problems, the expertise I did obtain had a marked impact on my findings. Without a fairly complete

understanding of the depletion allowance, for example, it would have been difficult to conclude with confidence that the change from 27½ percent to 22 percent had only minimal impact for the oil industry and that the industry could predict that impact. This was not the type of information industry officials were willing to offer in interviews and was something about which other respondents would or could only speculate. Yet it was vital to understanding industry behavior during the consideration of the 1969 tax bill, to knowing why small and large producers split on the issue, and to drawing a systematic picture of interest group behavior as was done at the end of Chapter 5. If we do not include policy analysis as a significant aspect of the political science discipline, the conclusions of our studies are more likely to prove invalid. Accepting that politics has something to do with the allocation of value or who gets what, etc., then we must understand who is getting what if our explanations of political behavior are to be sufficient.

A final implication of this study is a reconfirmation of interest groups as a worthy focus of political analysis. Since Bentley's original statement of "group theory," the study of interest groups has been an often maligned area of political science research. While it has gone through at least two major revivals since Bentley, one in the late 1920s and early 1930s and the other in the 1950s and early 1960s, it came under attack because the theory claimed more than was feasibly achievable from the analysis of group behavior and because the applications of the theory lacked linkage to it. In Chapter 1 this problem was more fully oulined. Yet the result has been that interest groups have received too little attention as a subject for political analysis and as a source of theory-building. The present study supports the view that interest groups are an important component of political life that can be understood in a systematic fashion. They not only have a definite impact on the outputs of political systems and subsystems, but they operate in ordered patterns that are affected by their political environment. In turn, interest group behavior indicates some things about the political environment (for example, where important decisions occur).

Naturally, this study is just an early step in building linkages between interest group theory and interest group behavior. There are several clear limitations in the present work that are of special interest. While the study answers some questions about the nature of interest group behavior, it opens up other areas of needed research. First, while the use of controls in this research did allow for some generalization, it clearly limited the scope of the analysis. It was possible to generalize about the nature of an interest group's behavior across a policy range in the Congressional process. But, as was noted in Chapter 3, the two issues selected here restricted the study of a group's activity to cases where it is in a defending position. The entire discussion of that factor indicates that the same group would be forced to adopt very different strategies and tactics if it were in a proponent position and its likelihood of success would be seriously decreased. A detailed analysis of proponent group activity is necessary to fill in major gaps in our knowledge of interest group behavior.

A second limit of this study is the policy range it covers. Neither issue enters a pure material politics level. Thus the extension of conclusions about interest group behavior at that level needs empirical confirmation. Nor is it entirely certain that the symbolic and material poles are the true ends of the policy continuum or are the only useful policy frameworks from which to analyze interest group behavior. The finding that interest group behavior is policy related merely opens the door to other work on policy-related variation.

A third area worthy of additional research involves the range of interest groups to which the findings of this study are applicable. The analysis here indicates that the behavior of mass public interest groups in the Congressional process may be very different from that of smaller private ones like the oil industry. For one thing, it is clear that the oil industry will usually benefit from privatized conflict where the important decisions are made at low visibility decision making points, while the former will prefer a high socialized conflict resolved at very visible decision making points. But this study was devoted in large part to the activities of smaller private interest groups; other types of groups received only auxiliary attention.

Finally, any study is subject to error even in its attempt to generalize about political behavior it claims to cover thoroughly. An attempt has been made here to keep this error minimal by controlling for possible nuisance variables or by accounting for them in the course of analysis. But as with any area of scientific investigation, some form of replication is desirable, be it replication in only the crudest sense, before the findings here are fully accepted.

Clearly, there remain many facets of interest group behavior open for study. Despite the attention they have received from political scientists, interest groups have not been fully explored. This study has, hopefully, forged some positive links between interest group behavior and theory. But of equal importance is the tone that this work tries to set for the systematic study of the role of interest groups in the policy process.

Appendix

Appendix

Methods and Limitations

The basis for this study comes from the examination of oil industry activity on water pollution and tax bills that came before Congress in the post-World War II era. I paid special attention to the tax bills of 1950, 1951, 1963, and 1969. Each of these involved some struggle over the depletion allowance. The water pollution bills of 1965, 1966, 1968, and 1969, which contained sections dealing directly with oil pollution, were also studied.

The Tax Reform Act of 1969 and the Water Quality Improvement Act of 1970,[1] among these bills, received the deepest consideration because they were the most recent pieces of legislation when I undertook this study and because they represent major breakthroughs in each issue area.

I relied on three basic sources of information for this study in addition to previous research on these issues: the public record; private information and records to which I gained access; and interviews with participants and observers in the decision making processes. The public record was the least difficult to obtain. I examined the committee hearings, committee reports, floor debates, and roll call votes in each issue area. In addition, I relied on *Congressional Quarterly's* reports on legislative action and news stories—primarily from *The Washington Post, The New York Times*, and *The Wall Street Journal*—for supplementary information.

These sources would clearly have been insufficient for my purposes. Much of what occurs in the Congressional process is not open to public inspection. Votes in committee may be far more meaningful than floor votes. Public hearings may be just a facade, with the important statements and decisions taking place in closed executive sessions. Thus, a major part of the investigation involved gathering information not available from the public record or from news accounts. This was somewhat more difficult to do. I was fortunate in obtaining access to several private records. I had the opportunity to examine the executive session transcripts of the House Public Works Committee during its mark-up of H.R. 4148 (the 1969 bill). It provided new information and confirmed some theories about the Committee's operation on the water pollution issue. More useful, however, were the files of a former Treasury official. Included in this were notes of Treasury meetings with oil industry officials, internal Treasury memos on the depletion issues, memos from Treasury to the White House on the Tax Reform Bill of 1969, and incomplete notes on the executive sessions of the House Ways and Means Committee and the Senate Finance Committee on that bill.

The most important sources of information were the interviews I conducted with approximately 50 individuals involved in these issues. Those interviewed

included members of the House and Senate, committee staff involved in the legislation, Executive agency officials, journalists, and lobbyists for the oil industry and other interest groups. Most of the interviews were conducted between January and August 1971. A few additional interviews were carried out in January 1972 and one in June 1972. They ranged in length from 20 minutes to three hours, averaging about an hour. I spoke with most respondents only once, preferring to talk with a new respondent with a similar perspective on the issue if additional information was needed. Only when a respondent was a unique source of information were second and sometimes third interviews held.

I faced two time limitations in interviewing. The interviewees often will cooperate only for short periods of time due to their schedules, and I could not conduct an infinite number of interviews.[2] Therefore, I adopted the following guidelines for the interview schedule. Respondents were asked first about the specific areas of the legislation and process they knew the best. Ways and Means members were questioned about the Committee's activities during the 1969 bill, how they got on the Committee, about oil industry activities in the House, and about comparisons with the Senate. Only when time allowed did we explore areas such as the oil industry's activity with the Treasury. Similarly Treasury officials were questioned primarily about industry activities there.

A second guideline for the interviews was to select individuals who were likely to hold differing views on the issue. This enabled me to obtain maximum information with a minimal number of interviews. For example, in interviewing members of the Ways and Means Committee about the change in depletion in 1969, I selected individuals who represented the different voting groups in the Committee in the executive session (See Chapter 2, Table 2-5).

Third, when I received similar answers to our questions from different perspectives and those answers also fit with the information obtained previously, I often stopped asking the questions and substituted others. When answers were opposed, I continued to ask the questions and probe for clarification.

It should be clear that there was no set questionnaire, although many of the respondents answered the same questions. The questions were all open-ended. Each interview was used as a building block for the next interview and not necessarily for survey purposes.

One additional comment should be made about the interviews. The subject matter was often so sensitive that I decided against using a tape recorder.[3] Often I took notes during an interview when it appeared not to interfere with the quality of the responses. When a respondent seemed particularly sensitive or uncooperative, I ceased to take notes. There was usually a marked improvement in the respondent's attitude when I put pen in pocket.

Clearly this technique, as with the others mentioned, has limitations. The major one is information loss. No doubt I forgot responses to some questions or remembered only parts of them. But I believe that some information would never have been volunteered had I insisted on recording interviews.

Notes

Notes

Notes–Chapter 1

1. Frank C. Porter, "House Unit Votes to Cut Off Exemption 18-7," *The Washington Post*, July 22, 1969, p.A:2.

2. Raymond A. Bauer, Ithiel de Sola Pool, and Lewis A. Dexter, *American Business and Public Policy* (New York: Atherton Press, 1963).

3. Theodore J. Lowi, "American Business, Public Policy, Case Studies, and Political Theory," *World Politics*, 16 (July 1964), pp. 617-715.

4. Bauer, op. cit., Second Edition (Chicago: Aldine-Atherton, 1972), p. ix.

5. Ibid.

6. E.E. Schattschneider, *Politics, Pressures and the Tariff* (New York: Prentice-Hall, Inc., 1935); Pendleton Herring, *Group Representation before Congress* (Baltimore: Johns Hopkins Press, 1929); Bertram M. Gross, *The Legislative Struggle: A Study in Social Combat* (New York: McGraw-Hill, 1953); Lester Milbrath, *The Washington Lobbyists* (Chicago: Rand McNally and Co., 1963); and Peter H. Odegard, *Pressure Politics: The Story of the Anti-Saloon League* (New York: Columbia University Press, 1928).

7. James Madison, "Federalist No. 10," and Alexis de Tocqueville, *Democracy in America* (New York: Colonial Press, 1869).

8. Arthur F. Bentley, *The Process of Government* (Bloomington: Principia Press, 1935). For a valuable retrospective of Bentley's contributions see Richard W. Taylor, ed., *Life, Language, Law* (Yellow Springs, Ohio: The Antioch Press, 1957). This collection of essays was done to honor Bentley. Of particular note are the title essay by Taylor and the essay by Charles B. Hagan, "The Group in Political Science."

9. Ibid., p. 204.

10. For discussion of the importance of falsifiability, see Karl R. Popper, *The Logic of Scientific Discovery* (London: Hutchinson, 1959), Chapter 4.

11. In fact, the main thrust of Bentley's book argued against collection and organization of political observations in terms of feeling and motivation and suggested substitution of groups for this task.

12. Odegard, op. cit.

13. Herring, op. cit.

14. Schattschneider, op. cit.

15. E.E. Schattschneider, *The Semisovereign People* (New York: Holt, Rinehart, and Winston, 1960), p. 21.

16. Ibid.

17. David Easton, *The Political System* (New York: Alfred A. Knopf, 1959), p. 176.

18. David B. Truman, *The Governmental Process* (New York: Alfred A.

Knopf, 1951). Truman's work was followed shortly thereafter by Earl Latham's *The Group Basis of Politics* (Ithaca, N.Y.: Cornell University Press, 1952) and "The Group Basis of Politics: Notes for a Theory," *American Political Science Review*, XLVI (June 1952).

19. Ibid., pp. 34-35.

20. Ibid., p. 352.

21. Giovanni Sartori, "Concept Misformation in Comparative Politics," *American Political Science Review* (December 1970), p. 1043. Recently economic theorists have tried to fill this void. In particular, the work of Mancur Olson, Jr. has attempted to distinguish certain organizational difficulties that make it difficult for mass membership interest groups to succeed. But Olson's theory is too simplified, and empirical evidence does not fit his claims. Mancur Olson, Jr., *The Logic of Collective Action* (New York: Schocken Books, 1968).

22. For a discussion of various policy typologies, see: Lewis A. Froman, "The Categorization of Policy Contents," in Austin Ranney (ed.), *Political Science and Public Policy* (Chicago: Markham, 1968), pp. 41-52.

23. Lewis A. Froman, "An Analysis of Public Policy in Cities," *Journal of Politics*, 29 (February 1967), pp. 94-108.

24. Lowi, op. cit.

25. Robert J. Salisbury, "The Analysis of Public Policy: A Search for Theories and Roles," in Austin Ranney (ed.), *Political Science and Public Policy* (Chicago: Markham, 1968); pp. 151-175.

26. Murray Edelman, *The Symbolic Uses of Politics* (Urbana, Ill.: University of Illinois Press, 1967).

27. Somehow, in considering the various policy frameworks on a rating scale involving a series of variables, Froman finds that his own scheme fares the best. Not only do I disagree with the overall ratings, but I find little justification for his evaluations.

28. Froman claims that city expenditures on welfare is an example of a segmental policy because "only a small segment of the population is involved, it affects different people at different times, and it is a continuing program." Froman, op. cit., p. 104. Yet a plausible case could be made for welfare expenditures' having impact on a substantial segment of a city and not just on the recipients.

29. If one pictured the four policy types as vectors then only four dimensions are involved. However, if continua are set up that connect two types, then six dimensions are required (a combination of four things taken two at a time or $\frac{4 \cdot 3}{2 \cdot 1} = 6$).

30. Edelman, op. cit., pp. 41-42.

31. Ibid., pp. 22-23.

32. Edelman, op. cit., p. 28.

33. John F. Manley, *The Politics of Finance* (Boston: Little, Brown, 1970)

and Stanley Surrey, "The Congress and the Tax Lobbyist—How Special Tax Provisions Get Enacted," *Harvard Law Review* (May 1957). Also see Philip M. Stern, *The Great Treasury Raid* (New York: Random House, 1962) and Joseph A. Ruskay and Richard A. Osserman, *Halfway to Tax Reform* (Bloomington, Ind.: Indiana University Press, 1970).

34. James L. Sundquist, *Politics and Policy* (Washington, D.C.: The Brookings Institution, 1968), pp. 322-381 and M. Kent Jennings, "Legislative Politics and Water Pollution Control," in Frederic N. Cleaveland (ed.), *Congress and Urban Problems* (Washington, D.C.: The Brookings Institution, 1969), pp. 72-109. Recently Charles O. Jones reviewed a broad selection of the literature on the environment and politics. His conclusion about its failings are similar to mine. He finds that only five of the fourteen books he examined offered "political knowledge on environmental issues" of any kind and that knowledge was of a limited nature. Seè Charles O. Jones, "From Gold to Garbage: A Bibliographic Essay on Politics and the Environment," *American Political Science Review*, 64 (June 1972), pp. 588-595.

35. Robert Engler, *The Politics of Oil* (New York: MacMillan, 1961).

36. We can examine variations between the two issues and within each issue over time.

37. Richard Fenno, *The Power of the Purse* (Boston: Little, Brown, 1964), pp. 505-506.

38. Lewis A. Froman, Jr., *The Congressional Process* (Boston: Little, Brown, 1967), pp. 13-15, and *Congressmen and Their Constituencies* (Chicago: Rand McNally, 1963), pp. 80-84.

39. John F. Manley, *The Politics of Finance* (Boston: Little, Brown, 1970), pp. 322-379.

40. Ibid., p. 279.

41. Warren Miller and Donald Stokes, "Constituency Influence in Congress," *American Political Science Review*, 57 (March 1963), pp. 45-56. Also see Julius Turner, *Party and Constituency: Pressures on Congress* (Baltimore: Johns Hopkins University Press, 1951), for an earlier treatment of constituency impact.

42. Duncan MacRae, *Dimensions of Congressional Voting: A Statistical Study of the House of Representatives in the Eighty-First Congress* (Berkeley: University of California Press, 1958).

43. David Mayhew, *Party Loyalty Among Congressmen* (Cambridge: Harvard University Press, 1966).

44. John Edgar Jackson, *A Statistical Model of United States Senators' Voting Behavior* (Cambridge: Ph.D. dissertation, Harvard University, 1968).

45. See Stephan A. Cobb, "Defense Spending and Foreign Policy in the House of Representatives," *Journal of Conflict Resolution*, Vol. 13, No. 3, pp. 358-369. James Clotfelter, "Senate Voting and Constituency Stake in Defense Spending," *Journal of Politics* (November, 1970), No. 4, pp. 979-983; Charles O.

Jones, "Representation in Congress: The Case of the House Agriculture Committee," *American Political Science Review*, LV (June 1961), pp. 358-367, and an unpublished paper by this author, "Model Cities: A Study of Urban Policy."

46. Lewis Anthony Dexter, "The Representative and His District," *Human Organization*, 16 (1947), pp. 2-13.

47. Charles O. Jones, "The Role of the Campaign in Congressional Politics," in M. Kent Jennings and L. Harmon Zeigler (eds.), *The Electoral Process* (New York: Prentice-Hall, 1966).

48. Warren Miller and Donald Stokes, "Party Government and the Saliency of Congress," *Public Opinion Quarterly*, 25 (Spring 1961), pp. 1-24.

49. Thomas Woodrow Wilson, *Congressional Government* (New York: Houghton-Mifflin, 1885), pp. 62-63.

50. Ibid., pp. 189-190.

51. See Fenno. op. cit.; Manley, op. cit.; Jones, op. cit.; James A. Robinson, *The House Rules Committee* (Indianapolis: Bobbs-Merrill, 1963); and James F. Murphy, "The House Public Works Committee" (Ph.D. dissertation, University of Rochester, 1969).

52. Frank J. Munger and Richard Fenno, Jr., *National Politics in Federal Aid to Education* (Syracuse: Syracuse University Press, 1962); and James L. Sundquist, *Politics and Policy* (Washington, D.C.: Brookings Institution, 1969) are excellent examples of this approach.

53. See Froman, *The Congressional Process*, op. cit., Nicholas A. Masters, "Committee Assignments," *American Political Science Review*, LV (June 1961), pp. 345-357; and Joseph S. Clark, *Congress: The Sapless Branch* (New York: Harper and Row, 1965).

54. Sundquist, op. cit., pp. 509-510.

55. Stanley S. Surrey, "The Congress and the Tax Lobbyist—How Special Tax Provisions Get Enacted," *Harvard Law Review*, LXX (May 1957), pp. 1145-1182.

Notes—Chapter 2

1. Some may argue that what are referred to here as independent variables may, in fact, only be contextual variables. This point is certainly a debatable one, especially in Chapter 3, where the nature of the committee handling the particular issue is treated as a source of variation. I would maintain, however, that committee behavior is a reduction of contextual and independent variables.

2. In fact, this rationale has been challenged on numerous occasions. Some of the arguments against it have been: (a) the risk in oil drilling is far less now than when depletion was adopted; (b) oil companies do not reinvest the depletion incentive in new exploration; and (c) the depletion allowance only works to assist those making profits, and by improving the tax situations of

those with profitable operations, it may actually discourage new firms from competing.

3. In 1969, for example, major U.S. oil companies paid 8.7 percent of their before tax earnings in federal income taxes. This was the highest rate since 1962. These figures were provided by *U.S. Oil Week* as inserted in the *Congressional Record* by Senator William Proxmire, October 27, 1971, pp. 16896-16898.

4. *Congressional Record*, February 22, 1944, p. 1959.

5. Ibid., February 23, 1944, p. 1965.

6. Hearings before the Ways and Means Committee, "Revenue Revision Act of 1950," 81st Congress, 2nd Session, Vol. I, p. 189.

7. Richard Fenno, *The Power of the Purse* (Boston: Little, Brown, 1966), p. 506.

8. "The Oil Producing Industry in Your State," (Washington, D.C.: Independent Petroleum Association of America, 1969).

9. To be sure to include states where there was a growing oil production or a residue of previous producing strength, the limited requirement of 1 percent of U.S. production during any one year was adopted.

10. While the amount or value of oil production may be a more meaningful figure than percentage of U.S. production, the percentage figure is acceptable. Data on total production was incomplete in some years and pricing problems make a dollar figure difficult to interpret. Moreover, if we look just outside this 1 percent level, we find states like Kentucky. Crude petroleum production there has been on a steady decline and is now around fifteen million barrels a year. A similar situation exists in Pennsylvania. In neither state has the petroleum industry been a major one in recent decades, except in a few counties which account for most of the production.

11. *The 1966 Mineral Yearbook* was selected here because it had more complete information. Some states provided Interior with data on all three items. Often I had to rely on the Yearbook's discussion of a given county in deciding whether significant production was taking place.

12. David Truman, *The Governmental Process* (New York: Alfred Knopf, 1951), p. 322. Truman provides the following case, which indirectly supports our point. "Take as an example the provision for equal representation of states in the Senate. . . . This has allowed agricultural interest groups that are predominant in many thinly populated states more points of access in the Senate than urban groups whose members are concentrated in a few populous states." What Truman fails to note is that we would expect that urban groups would fare better in the House than in the Senate and agricultural groups better in the Senate than in the House. The important point, however, is that distribution of the interest has an affect on where it is likely to do better. Unfortunately, our example does not meet the "proof of the pudding." Urban groups do best in the Senate and agriculture groups best in the House. Other factors confound the expected relationship.

13. The 27½ percent figure was arrived at in the conference committee, where conferees agreed to split the difference between the House and the Senate bills. The logic of this figure has consistently been a point of attack for depletion opponents. They claim that the figure is not based on any notion of the actual difficulty in finding new sources of oil. Almost every source I have consulted claims that the Senate had voted a 30 percent figure and the House 25 percent, and they agreed in conference to split the difference. In checking the *Congressional Record* on this subject, I have found that this was not exactly the case. The 25 percent figure did not appear in the House bill, while the 30 percent figure was in the Senate bill. As best as I can reconstruct, the 25 percent figure was the bargaining position taken by the House conferees.

14. Included here were aplite, wollastonite, magnetite, dolomite, brucit, sodium chloride, calcium chloride, magnesium chloride, potassium chloride, and bromine.

15. For a discussion of Gamma see L.A. Goodman and W.H. Kruskal, "Measures of Association for Cross-classifications," *Journal of the American Statistical Association*, 49, 1954, pp. 732-764. Gamma will equal 1.0 when any cell in the table is 0.

16. Paul Douglas, "The Problem of Tax Loopholes," *The American Scholar*, Winter 1967-1968, Vol. 37, No. 1, pp. 29-30. For an account of Douglas' struggle against depletion and for an explanation of why he chose depletion instead of other tax provisions such as intangible drilling expensing, see his autobiography, *In the Fullness of Time*, (New York: Harcourt, Brace, Jovanovich, 1971), pp. 428-433.

17. For a discussion of ways Senators may perceive what interests are part of their constituencies see my paper, "Senators' Constituencies: A Redefinition," a paper delivered at the 1971 American Political Science Association Convention. Douglas' view of the oil industry in Illinois fits remarkably well with my redefinition.

18. Robert Engler, *The Politics of Oil* (New York: Macmillan, 1961).

19. *Congressional Record*, August 11, 1958, p. 16900.

20. Ibid.

21. Letter from Peter Dominick.

22. Ronnie Dugger reports that Moss was offered a large campaign contribution in exchange for support for depletion when he first ran for the Senate in 1958. The 1960 vote may have been an attempt on Moss' part to show his independence. He has been a consistent supporter of depletion since that vote. See Ronnie Dugger, "Oil and Politics," *The Atlantic* (September 1969), p. 74.

23. *Congressional Record*, June 28, 1954, p. 9306.

24. Ibid., p. 9307.

25. Ibid., June 25, 1959, p. 11927.

26. Ibid., April 19, 1967, p. 10193. Harris' public position regarding special tax provisions changed markedly once he announced his intention to retire from

the Senate and started his "populist" campaign for the presidency. His constituency was no longer limited to Oklahoma.

27. Hearings before the Ways and Means Committee, "Revenue Act of 1951," 92nd Congress, 2nd Session, 1951, p. 101.

28. A former Treasury official reports that Boggs was just carrying out his part of an agreement with Vanik. Vanik agreed not to go for a bigger cut if Boggs would accept 20 percent. In fact, this respondent believes that the Gibbons motion was just for show. It nevertheless separates the most adamant of the depletion opponents from the rest of the Committee.

29. See John F. Manley, *The Politics of Finance* (Boston: Little, Brown, 1970). Manley discusses the relationship between members' evaluation of their committee and its level of autonomy.

30. Froman, op. cit., p. 24.

31. Ways and Means Hearings, 1951, op. cit., p. 1537.

32. Ibid., p. 1640.

33. *Congressional Record*, June 21, 1951, p. 6965.

34. *Congressional Record*, September 28, 1951, pp. 12320-12322.

35. Ibid., p. 12323.

36. Ibid., June 20, 1951, p. 11742.

37. Ibid., August 11, 1958, p. 16900.

38. E.E. Schattschneider, *The Semisovereign People*, (New York: Holt, Rinehart and Winston, 1960).

39. Ways and Means, 1951, op. cit.

40. *The New York Times*, June 29, 1957, p. 9:2.

41. Bernard D. Nossiter, "Ex-Treasury Chief Received Oil Funds," *The Washington Post*, July 16, 1970, p. A:1.

42. Connally's ties to the oil industry are well known. Most often cited among these is his work for, and close association with, the late Mr. Sid Richardson, a multimillionaire Texas oilman.

43. Peter Milius, "Connally Cool to Tax Reform Now, Opposes 20 Percent Social Security," *The Washington Post*, February 29, 1972, p. 72:3.

44. The 1963 tax bill amended the provision allowing oil companies to group properties before computing depletion. Previously companies had grouped profitable and unprofitable holdings to avoid reaching the 50 percent of net limitation on their high profitable wells. This made it possible for large companies to realize a higher percentage depletion than small concerns that did not have enough properties to manipulate groupings. See Philip M. Stern, *The Great Treasury Raid* (New York: Random House, 1964), for a simplified explanation of the property groupings provision.

45. A recent conversation with the Treasury official responsible for proposing the rule change revealed that it was not his original intention to go after depletion. His efforts were directed at establishing a computation method for the posted prices on domestic petroleum, the price at the well that the refinery

must pay for crude oil. He wanted to know whether they were fair, but without a standardized pricing method this was impossible. However, this official realized that it would also question the amount allowed for depletion. Oil officials were quick to respond. The official who initiated the proposal was called into his boss' office and told, "These people are just too powerful. You have to give them what they want." The official also reports that Stanley Surrey received a call from Russell Long charging Treasury with abrogating legislative responsibility.

46. Surrey had authored several articles on tax reform prior to assuming his position at Treasury; one highly critical of the potential for tax reform coming from Congress is probably his best. "The Congress and the Tax Lobbyist—How Special Tax Provisions Get Enacted," *Harvard Law Review* (May 1957), pp. 1145-1182.

47. Manley finds this is a commonplace occurrence. He quotes one Treasury official on the consultants it hires to write its proposals, "Just take a look at the provisions and who they affect and you can be pretty sure we've met with the major spokesmen." Manley, op. cit., p. 357.

48. Manley, op. cit., p. 377.

49. For numerous examples of this occurrence, see Murray Edelman, *The Symbolic Uses of Politics* (Urbana: University of Illinois Press, 1967).

50. Alexander Heard, *The Costs of Democracy* (Chapel Hill: University of North Carolina Press, 1960), pp. 60-65. Heard documents the case of interests who "cover their bets" and contribute to both major parties.

51. Alexander, op. cit., p. 184.

52. Emile B. Adler, "Why the Dixiecrats Failed," *Journal of Politics*, Vol. 15, (1953), pp. 365-366.

53. "Lyndon Johnson and Oil," *Parade Magazine*, November 10, 1958.

54. Ibid.

55. Reported in *The New York Times*, November 2, 1968, p. 1:2, and November 3, 1968.

56. Murray Segger, "Depletion Stand Hurts Humphrey," *Los Angeles Times*, February 16, 1969, p. 19.

57. James Tanner and Norman Perlstine, "Oilmen Expect Better Treatment by Nixon than They Got from his Predecessors," *Wall Street Journal*, November 20, 1968. Reprinted from *The Wall Street Journal,* © Dow Jones & Company, Inc., 1968.

58. Drew Pearson and Jack Anderson, "Washington Merry-Go-Round," *The Washington Post*, May 7, 1969, p. B15.

59. Rowland Evans, Jr., and Robert D. Novak, *Nixon in the White House: The Frustration of Power* (New York: Vintage Books, 1972), pp. 218-222. Later, at a press conference on September 26, Nixon explained that he still supported depletion, but that he was a "political realist."

60. Truman, op. cit., p. 324.

61. Thomas F. Field, "How Tax Laws are Changed—How the Process Can Be Improved," speech to National Tax Association Annual Conference, September 21, 1970, pp. 4-5.

62. Manley, op. cit., Chapter 7.

63. Prior to the 89th Congress water pollution bills were regularly subjected to recommital votes in the House. These were strong party line votes and did not demonstrate a relationship between constituency ties and voting behavior. Further, none of these bills were focused at the oil industry directly, as was the case with legislation from 1965-1969. For examination of these early votes, see M. Kent Jennings, "Legislative Politics and Water Pollution Control 1956-1961," in Frederick N. Cleaveland (ed.), *Congress and Urban Problems* (Washington, D.C.: The Brookings Institution, 1969), pp. 72-109.

64. *Congressional Record*, October 11, 1968, pp. 31108-31109.

65. *Congressional Quarterly Almanac*, 1968, pp. 569-575.

66. In 1967 the subcommittee also had control but held only oversight hearings and produced no bill.

67. Much of the basis of our discussion of the House Public Works Committee comes from James T. Murphy, "The House Public Works Committee" (Ph.D. Dissertation, University of Rochester, 1969). Murphy devotes a section to the reasons for growth of subcommittee autonomy under the weak chairmen who followed Congressman Whittington (Democrat, Mississippi) as leader of the Public Works Committee.

68. Ibid., p. 6.

69. Rundall B. Ripley, "Congress and Clean Air: The Issue of Enforcement, 1963," in Cleaveland (ed.), op. cit., pp. 224-278.

70. In 1945 four water pollution control bills were introduced in identical forms in the House and Senate. None were reported out of either the Senate Commerce Committee (then responsible for pollution legislation) or the House Public Works Subcommittee on Rivers and Harbors.

71. *Congressional Record*, June 14, 1948, p. 8197.

72. Murphy, op. cit.

73. Ibid., p. 233.

74. Hearings before the Committee on Public Works, House of Representatives on HR 4148 and Related Bills, 91st Congress, 1st Session, p. 96.

75. Ibid.

76. Ibid., pp. 352-353.

77. Hearings before the Committee on Public Works, House of Representatives on HR 15906 and Related Bills, 90th Congress, 2nd Session, p. 79.

78. Donald Matthews, *U.S. Senators and Their World* (Chapel Hill: University of North Carolina Press).

79. Jennings, op. cit., p. 74.

80. Ibid., p. 83.

81. Ibid., p. 85.

82. Murphy, op. cit., p. 41.

83. David Zwick and Marcy Benstock, *Ralph Nader's Study Group Report on Water Pollution—Water Wasteland* (New York: Crossman Publishers, 1971), p. 58. Admittedly, the report did not deal with oil pollution specifically but with more general water pollution problems.

84. In 1965 the Texas State Legislation removed certain regulatory powers regarding oil pollution from the pollution board and transferred them to the Texas Railroad Commission—a body closely tied to the oil industry. See "Texas Moves to Keep Right to Regulate Pollution," *The New York Times*, December 5, 1965, p. 166:3.

85. Zwick, op. cit., p. 60.

86. Ibid., pp. 62-64.

87. Commissioner Dominick left his position in late 1970 for a promotion to head the new pesticides division at EPA. Thus, the turnover in leadership continues.

88. Hearings before the Committee on Public Works, House of Representatives on the Water Quality Control Act, 89th Congress, 1st Session, p. 351. Section 5 would have given the Secretary of Health, Education, and Welfare the power to set water quality standards.

89. "Polluters Sit on Antipollution Boards," *The New York Times*, December 12, 1970, p. 1:1. A year later *The Times* reported that the situation had improved. It claimed that the number of states with pollution boards had dropped from 35 to 32. "States Curtailing Polluters on Pollution Control Units," *The New York Times*, December 19, 1971, p. 1:2.

90. On the floor votes on the Tax Reform Act of 1969, however, Morton voted against recommitting the bill.

91. Hearings before the Committee on Public Works, House of Representatives on HR 4148, op. cit.

Notes—Chapter 3

1. This point is made elsewhere in the legislative process literature. See David B. Truman, *The Governmental Process* (New York: Alfred A. Knopf, 1951), p. 330 and Lewis A. Froman, Jr., *The Congressional Process: Strategies, Rules and Procedures* (Boston: Little, Brown, 1967).

2. See Raymond E. Wolfinger, "Filibusters: Majority Rule, Presidential Leadership, and Senate Norms," in Wolfinger (ed.), *Readings on Congress* (Englewood Cliffs, New Jersey: Prentice-Hall, 1971), pp. 286-305.

3. For accounts of these legislative struggles see James A. Sundquist, *Politics and Policy* (Washington: The Brookings Institution, 1968); Richard F. Fenno, "The House of Representatives and Federal Aid to Education," in Robert L. Peabody and Nelson W. Polsby, eds., *New Perspectives on the House of*

Representatives (Chicago: Rand McNally, 1963), pp. 195-235, and Froman, op. cit., p. 93.

4. The three prime examples of leadership stacking committees are the expansion of the Rules Committee in 1961, the appointment of pro-Medicare members to Ways and Means in the early 1960s, and the liberal ideological make-up of the House Education and Labor Committee. For the first two of these, see Milton C. Cummings, Jr. and Robert L. Peabody, "The Decision to Enlarge the Committee on Rules," in Polsby and Peabody, op. cit., pp. 167-194; and John F. Manley, *The Politics of Finance* (Boston: Little, Brown, 1970), pp. 27-29.

5. This is more the case in the Senate, where party leadership has nearly complete control of scheduling than in the House. See Froman, op. cit., pp. 110-127 and Randall B. Ripley, *Majority Party Leadership in Congress* (Boston: Little, Brown, 1969), pp. 1-19.

6. In the 81st Congress eight bills reached the floor through use of a 21-day rule and in the 89th Congress six bills came to the floor in this manner. The former included: anti-poll tax, minimum wage, and housing bills, and the latter included: repeal of section 14b of Taft-Hartley, school construction, and employment discrimination among others. See Froman, op. cit., pp. 97-99, and James A. Robinson, *The House Rules Committee* (Indianapolis: Bobbs-Merrill, 1963), Chapter 4.

7. Froman, op. cit., Chapter 11.

8. Peter Bachrach and Morton Baratz, *Power and Poverty* (New York: Oxford Press, 1970), p. 44.

9. Ibid., p. 17.

10. If all decisions occurred at once, the resources would be greater. When decisions are sequential, time and staff resources, for example, need not be as great. Nevertheless, defending groups can cultivate on decision point continuously even when decisions are not being made there, while proponent groups need to jump along to each new decision spot as the issue moves.

11. *Revenue Act of 1951*, Hearing on H.R. 4473 before the Finance Committee of the U.S. Senate, 82nd Congress, 1st Session.

12. Ibid., pp. 106-107.

13. *Revenue Act of 1963*, Hearings on H.R. 8363 before the Finance Committee of the U.S. Senate, 88th Congress, 1st Session, p. 212.

14. According to a member of the White House staff under President Johnson, Hartke and McCarthy were considered safe on depletion issue because Russell Long had arranged financial support from oil interests to assist their Senate campaigns.

15. *Congressional Quarterly Almanac 1969*, p. 620.

16. This is the case when a House bill includes a provision excluded from a Senate bill or vice versa; a successful floor amendment may achieve the same purpose. The latter are rare in the issue areas we are considering, and the former

may require a conference committee, which, as we shall see, does not often keep the additional provisions.

17. Fred Graham, "Oil Pollution Act is Found Crippled," *The New York Times*, April 16, 1967, p. 41:1.

18. Ibid.

19. *Congressional Quarterly Almanac 1968*, pp. 569-575.

20. *Federal Water Pollution Control Act Amendments—1969*, Hearings before the Committee on Public Works, House of Representatives, 91st Congress, 1st Session, on H.R. 4148 and Related Bills.

21. There are some rare exceptions. Senator John Williams of Delaware was a leading opponent of depletion allowances. It should be noted that Delaware ranks 50 among the states in mineral production. On the House Public Works Committee one Republican, Representative John Baldwin of California, is remembered as siding with some Democrats in favoring strong water pollution control measures.

22. Manley, op. cit., pp. 26-27.

23. Ibid., pp. 48-49.

24. Until the 1963 hearings, Herman Eberharter (Dem., Pa.) was the only member of Ways and Means who had asked questions or made statements during the hearings that could in any way be construed as anti-depletion.

25. While this is a perception, it is an incorrect one. At least, one member of the Committee went on for expressly this purpose.

26. Vanik claims that Rayburn did block him earlier.

27. Manley, op. cit., p. 48.

28. Manley, in discussing this point, was interested in members' perceptions as to how they got on the Committee and not with analyzing when they got on.

29. Manley reports that Herlong was "a shade too conservative for Rayburn's tastes" but was too popular among Democrats for him to be denied the vacancy by either Rayburn or Carl Vinson (Dem., Ga.), who wanted a Georgian to fill the open spot on Ways and Means. Manley, op. cit., p. 34.

30. Peter Bachrach and Morton Baratz, "Two Faces of Power," *American Political Science Review* (Vol. LVI, 1962), pp. 947-952.

31. Two newspaper reporters, Donald L. Bartlett of *The Cleveland Plain Dealer* and Murray Seegar of *The Los Angeles Times*, each wrote numerous articles on depletion. Our interviews reveal that they were being fed information on Ways and Means activity from certain Committee members.

32. Rowland Evans and Robert Novak, *Lyndon B. Johnson: The Exercise of Power* (New York: Signet, 1969), pp. 113-114.

33. Frank V. Fowlkes and Harry Lenhart, Jr., "Two Money Committees Wield Power Differently," *The National Journal*, April 10, 1971, p. 796.

34. Paul H. Douglas, *In the Fullness of Time* (New York: Harcourt Brace Jovanovich, Inc., 1921), p. 427.

35. Ibid.

36. The most vehement case against the conservative nature of the Steering Committee was made by Joseph S. Clark in *The Senate Establishment* (New York: Hill and Wang, 1963).

37. For an extensive comparison of Johnson and Mansfield in the majority leadership position see John G. Stewart, "Two Strategies of Leadership: Johnson and Mansfield" in Nelson Polsby (ed.), *Congressional Behavior* (New York: Random House, 1971), pp. 61-92. Stewart claims " . . . Mansfield relinquished almost all personal control and influence over the Steering Committee's decisions," pp. 73-74.

38. *National Journal*, op. cit., p. 796. Bentsen has since been appointed to the Finance Committee at the start of the 93rd Congress.

39. Stephen Horn, *Unused Power* (Washington: The Brookings Institution, 1970), p. 24

40. Manley, op. cit., pp. 27-29.

41. One of the four Democratic vacancies was caused by the defeat of Paul Douglas for re-election to the Senate in 1966. This vacancy did not allow for increasing the number of anti-depletion members. Since 1969 there have been four Democratic vacancies on Finance. At the beginning of the 92nd Congress, Nelson of Wisconsin and Mondale of Minnesota replaced Gore of Tennessee and McCarthy of Minnesota, and at the beginning of the 93rd Gravel of Alaska and Bentsen of Texas succeeded Anderson of New Mexico and Harris of Oklahoma. This maintained and perhaps even improved oil industry strength on the Finance Committee.

42. James T. Murphy, "The House Public Works Committee" (Ph.D. Dissertation, University of Rochester, 1969), p. 39.

43. David W. Rohde and Kenneth A. Shepsle in a recent article show how variability in types of Congressmen who request the same committee assignment will affect who receives the assignment. They show various factors are related to success in getting committee assignment. See David W. Rohde and Kenneth A. Shepsle, "Democratic Committee Assignments in the House of Representatives: Strategic Aspects of a Social Choice Process," *American Political Science Review*, 67 (September 1973), No. 3, pp. 889-905.

44. The five are Jim Wright (Tex.), Ed Edmondson (Okl.), Frank Clark (Pa.), T. Ashton Thompson (La.), and Ken Gray (Ill.).

45. This has insured representation of an oil district on the conference committees. Most frequently the oil district conferee has been Jim Wright. He has served as a conferee on all water pollution bills since 1965.

46. David Price, *Who Makes the Laws?* (Cambridge, Mass.: Schenkman Publishing Co., 1972), p. 177. Price's point is well taken, but even if these eight members focused attention elsewhere, it still leaves six with whom Long had to cope. And when it comes to a major tax bill, even those with other responsibilities often get involved. Moreover, compared to other Senate committees and subcommittees, this level of attention may still be high.

47. The lack of subcommittees, however, does provide the chairman with additional power. He and the ranking minority member control the committee's staff. With no subcommittees, there are no subcommittee chairmen with independent staffs. This may tend to draw the attention of some Finance members to lesser committees where they are subcommittee chairmen and away from Finance. In turn, this further increases the control of the chairman. Still, in 1969 a substantial number of the Finance members were involved with the Tax Reform Act. In 1971, Finance added a standing subcommittee on Veteran's Legislation in an attempt to stop formation of a Senate Committee on Veterans Affairs. See Price, op. cit., p. 177. At the beginning of the 93rd Congress, Finance decentralized further through establishment of six subcommittees.

48. Among the norms of behavior cited as proper for freshmen Senators is that of apprenticeship, which according to Matthews includes: subordinate status, keeping his mouth shut, and showing respect for elders. Donald R. Matthews, *U.S. Senators and Their World* (Chapel Hill: University of North Carolina Press, 1960), p. 93.

49. At the start of the 92nd Congress the Muskie Subcommittee expanded to thirteen members. For all practical purposes, it could be considered the same as the full Public Works Committee—which had seventeen members. The major exception was that Muskie chaired the former and Randolph chaired the latter. This, however, may indicate that Muskie will not have as free a hand in the future in designing pollution control legislation. By comparison, in 1965 this subcommittee had only nine members.

50. In his article, "Committee Assignments in the House of Representatives," Nicholas A. Masters claims that no members of the House can hold two semi-exclusive committee assignments at the same time. There are an ample number of exceptions to this rule. On Public Works alone, there were three violations of the rule: Miller (Rep., Ohio) served on Agriculture and Henderson (Dem., N.C.) and Olsen (Dem., Mont.) served on Post Office and Civil Service. See *American Political Science Review*, 55 (June 1961), pp. 345-357. For a valuable perspective of how attention to committee work varies across a range of House committees by the motivations of the members, among other factors, see Richard Fenno's *Congressmen in Committees* (Boston: Little, Brown, 1973). While Fenno does not deal with Public Works specifically, the discussion here merges nicely with the boundaries he establishes.

51. Murphy, op. cit., p. 193.

52. The final wording called for federally "promulgated" standards as opposed to federally "approved" standards. Ibid., p. 286.

53. Ibid., p. 9.

54. Ibid., p. 6.

55. David Zwick and Marcy Benstock, *Water Wasteland* (New York: Grossman, 1971), p. 428.

56. Ibid., p. 429.

57. Ibid., p. 83.

58. Raymond A. Bauer, Ithiel de Sola Pool, and Lewis A. Dexter, *American Business and Public Policy* (New York: Atherton Press, 1963). They found that lobbyists tended to work only with legislators friendly to their position and to avoid those they perceived as hostile.

59. Fenno, op. cit.

60. National Journal, op. cit., p. 788.

61. Ibid., p. 788.

62. Some respondents claim that this was the reason for the surprising Finance vote on restoration to 27½ percent.

63. The one exception was in 1963, when the Committee accepted one part of the Kennedy Administration reform proposals on depletion. The proposal made the "property grouping" regulation more restrictive. Prior to this, companies could group non-contiguous properties before computing depletion. This allowed companies to group highly profitable properties with highly unprofitable ones and receive the full amount of depletion (i.e., not exceeding the 50 percent of net limitation before reaching the 27½ percent of gross income allowed for depletion). Property grouping was limited to contiguous properties by the 1963 legislation.

64. Original motivation for change in the depletion allowance came from a list of tax reforms prepared at Treasury under Secretaries Joseph Barr and Henry Fowler. In December 1968 they urged outgoing President Johnson to send a tax message to Congress. (See "Tax Reformers Renew Assault on Depletion," by Murray Seegar, *Los Angeles Times*, December 29, 1968.)

65. In a speech given in Houston on September 6, 1968, Nixon is quoted as saying: "As a Congressman 21 years ago, as a Senator, as Vice President of the United States, as a candidate for President in 1968, I opposed the reduction of the depletion allowance because I want these great resources developed in Texas and across this nation." Humphrey refused to make this commitment. See *The Los Angeles Times*, February 16, 1969, Section A, p. 6.

66. Plowbacks would allow depletion to the extent that the taxpayer "expended an equivalent amount for exploration or development" of a natural resource. According to Evans and Novak, the plowback memo was never seen by the President. "It went to John Erlichman and there it stopped." But this only refers to a memo drafted in late June and does not cover the July 17th memo. Rowland Evans and Robert D. Novak, *Nixon in the White House: The Frustration of Power* (New York: Vintage Books, 1971), p. 220.

67. David Price provides additional support for this refinement of the appeals court hypothesis. He claims that, "Groups would naturally view the Senate Committee as a place for last-ditch attempts at obstruction or amendment, rather than a likely point for the development of new or alternative proposals." Price, op. cit., p. 187.

68. E.E. Schattschneider, *The Semisovereign People* (New York: Holt, Rinehart and Winston, 1960).

69. Manley, op. cit., pp. 220-221.

70. From April 1, 1932 until February 27, 1973 every tax bill considered on the House floor was covered by a closed rule. The streak was broken when the House Rules Committee granted an open rule on H.R. 3577, a bill extending the interest equalization tax. Interestingly, no floor amendments were offered although they were permitted. See *CQ Weekly Report*, Volume XXXI, No. 9, March 3, 1973, p. 465.

71. Manley, op. cit., pp. 220-221.

72. Ibid., p. 236.

73. While the Senate floor is a relatively visible decision point, most members of the public still do not pay attention to it. The bulk of amendments accepted there continue to improve the position of private, organized groups. These groups are often ones who did lose on the House side and appealed to the Senate. However, the reason for the lack of public attention may be due more to the nature of tax legislation than to the visibility. This point will be further developed in Chapter 4.

74. Intangible drilling costs allow oil companies to expense labor and other drilling costs of a well rather than capitalizing them and depreciating them as an asset.

75. Charles Lindblom, "Decision Making: 'The Science of Muddling Through,'" *Public Administration Review*, 19 (April 1959), pp. 79-88, and *The Intelligence of Democracy* (Glencoe: The Free Press, 1966), and Aaron Wildavsky, *The Politics of the Budgetary Process* (Boston: Little, Brown, 1964).

76. The Treasury Department has required oil companies to supply information on production, drilling, and investment on Form 0, which was drawn up by Treasury. Before 1971 information supplied on Form 0's was supposed to be public, the Treasury Department has not made it available (although it was requested, according to former Treasury officials).

77. Murphy, op. cit.

78. For more extensive analysis of the committees see Manley, op. cit., Murphy, op. cit., and Price, op. cit.

79. Ibid.

80. Ibid.

81. Norms of behavior assist the situation—especially "reciprocity." For a discussion of the concept see Matthews, op. cit., and Fenno, op. cit.

Notes—Chapter 4

1. E.E. Schattschneider, *The Semisovereign People* (New York: Holt, Rinehart and Winston, 1960), p. 2.

2. Stanley Surrey, "The Congress and the Tax Lobbyist—How Special Tax Provisions Get Enacted," *Harvard Law Review* (May 1957), p. 1164. Copyright 1957 by the Harvard Law Review Association.

3. Mancur Olson, Jr., *The Logic of Collective Action* (New York: Schocken Books, 1968).

4. Ibid., p. 165.

5. Ibid., pp. 133-135.

6. Hearing before the Ways and Means Committee, "Revenue Revision Act of 1950," U.S. House of Representatives, 81st Congress, 2nd Session, p. 744.

7. Ibid., pp. 2632-2633.

8. "Revenue Act of 1951," Hearings before the Ways and Means Committee on H.R. 4473, U.S. House of Representatives, 82nd Congress, 1st Session.

9. Ibid., pp. 432, 890.

10. *House Report No. 586*, 80th Congress, 2nd Session, pp. 29-30.

11. Thomas E. Field, "How Tax Laws are Changed, How the Process Can be Improved," Speech to the National Tax Association, September 21, 1970, p. 11.

12. Surrey, op. cit., p. 1166. Copyright 1957 by the Harvard Law Review Association.

13. *The Washington Post*, February 29, 1972, p. A 2:3.

14. Surrey, op. cit., p. 1165. Copyright 1957 by the Harvard Law Review Association.

15. John F. Manley, *The Politics of Finance* (Boston: Little, Brown, 1970) and Surrey, op. cit.

16. Surrey, op. cit., p. 1166. Copyright 1957 by the Harvard Law Review Association.

17. The six were George Mahon (Dem., Texas), O.C. Fisher (Dem., Tex.), Graham Purcell (Dem. Tex.), Richard White (Dem., Tex.), Page Belcher (Rep., Okl.), and Wayne Aspinall (Dem., Col.).

18. The five were Joelson (Dem., N.J.), Andrew Jacobs (Dem., Ind.), Joseph Minish (Dem., N.J.), Lloyd Meeds (Dem., Wash.), and Mario Biaggi (Dem., N.Y.).

19. Consad 1969: Consad Research Corporation, "The Economic Factors Affecting the Level of Domestic Petroleum Reserves" (Pittsburgh: Consad, 1969). The Consad Report became one part of *The Tax Reform Studies* initiated at Treasury under the Johnson administration.

20. "Hearings before the Committee on Ways and Means, House of Representatives, 91st Congress, 1st Session on the Subject of Tax Reform" (Washington: U.S. Government Printing Office, 1969), pp. 3388-3426.

21. Ibid., p. 3427.

22. Since Mills and Byrnes were chairman and ranking minority member respectively, it is not totally surprising that they did much of the questioning. But it is the first time that either of them expressed a significant interest in depletion. Also, as Manley describes it, Ways and Means allows for active participation of juniors but this is often confined to executive sessions. "In public hearings most of the good questions may be gone by the time they get to the junior members but in executive sessions the process is more fluid." John F. Manley, *The Politics of Finance* (Boston: Little, Brown, 1970), p. 95. Conable who joined the Committee in 1967 is an exception. He has been an active

participant despite his junior status. Part of this stems from Conable's abilities, which other members and staff regularly mention, and part from the lesser abilities of the Republican members between Byrnes and Conable. With Byrnes' retirement at the end of the 92nd Congress, many speculated that Conable would become the predominant force among the Committee Republicans.

23. Byrnes' shock here was due to the fact that Atlantic's operations were almost totally domestic. Companies with foreign operations usually pay low federal taxes because they can credit foreign income taxes against U.S. taxes.

24. Ways and Means, 1969, op. cit., p. 3199.

25. Ibid., p. 3200.

26. Ibid., p. 3204.

27. David Truman, *The Governmental Process* (New York: Alfred Knopf, 1951), p. 34. One may differ with the notion that taxpayers were only a potential group. Certainly, the taxpayers' revolt, while diffuse, was nevertheless felt on Capitol Hill in terms of such things as the mail Congressmen received. But the term "potential group" still applies since there is no sense in which interest provided the basis of interaction among individuals. Truman, op. cit., p. 34.

28. This does not mean we accept the overly optimistic pluralist conclusions that Truman elaborates. In fact, as we shall discuss in the next chapter, these actions led the Committee to take steps to quiet the sources of the demands rather than to meet the demands.

29. United States Department of the Interior, *Mineral Yearbook* 1968 (Washington: D.C.: Government Publishing Office, 1969).

30. Burleson's views were already known to Ways and Means since he was a member of the Committee.

31. Michael Barone, et al., *The Almanac of American Politics* (Boston: Gambit, 1972), p. 657.

32. Obviously our measure of a district with an oil constituency is not sufficient for more exact purposes. Since it is only nominal measurement, it does not allow us to formally rank oil districts by strength of this interest. But examination of the districts represented by those who testified in 1969 compared to those who testified in 1950 and 1951 leads us to believe that this is a reasonable conclusion.

33. Some of the reasons for this relate to economies of scale. Others come directly from advantage in the tax code for broader drilling operations. These include offsetting loss and gain on contiguous properties and use of foreign oil operations that increase profitability.

34. IPAA does not, as some people think, just represent small, independent producers. Some of IPAA's members, such as Murphy Oil, are in a size category with the majors who belong to API.

35. Ways and Means, 1969, op. cit., p. 3181.

36. Ibid., p. 3188.

37. Ibid., p. 3255.

38. Ibid., p. 3273.

39. Ibid., p. 3282.

40. Ibid.

41. "Hearings before the Finance Committee on H.R. 13270," United States Senate, 91st Congress, 1st Session, Part 5, October 1, 1969, p. 4493.

42. After losing the fight to restore depletion fully, Russell Long moved that the Finance Committee vote to raise the net limitation to 65 percent, as he put it, to "help the little guy."

43. The Proxmire proposal called for depletion to remain at 27½ percent for the first $5 million of gross income, 21 percent for the second $5 million, and 15 percent for gross income above that. Ibid., p. 4210.

44. Brookings Institution luncheon, January 12, 1971.

45. Fannin's question.

46. Joseph Spear, "Speak Up, There's a Lobbyist in My Ear," *The Washington Monthly*, April, 1972, p. 19. The entire exchange is far too long to produce, but this sample gives a feeling of the hostility level. The Spear article juxtaposes Stanton's treatment with that afforded industry lobbyists. The article appeared some three to four months after I interviewed Stanton. At that earlier time he related the happenings to me.

47. Olson, op. cit., pp. 132-135.

48. Robert Lane, *Political Life* (Glencoe, Illinois: Free Press, 1959).

49. This is rapidly ceasing to be the case. Proof of pollution in itself has become complex, requiring witnesses of collection of effluents and qualitative and quantitative measurement of contents. In addition, the technology for pollution control has become increasingly intricate, requiring new levels of expertise to evaluate the effectiveness of control equipment and procedures. However, none of this interferes with an individual's being able to see oil on water or black smoke coming from a smokestack.

50. *Congressional Record*, October 8, 1969, p. 28959.

51. Ibid., pp. 28960-28963.

52. Ibid., p. 29098.

53. Barone, op. cit., p. 66.

54. The League is regularly cited as the exception from the other older groups working on pollution matters. It is credited by all parties as being one of the few groups that monitors the entire process. In addition, with its highly decentralized structure the League is able to keep track of enforcement in local areas and to feed information to the national headquarters.

55. Support for this view comes from James T. Murphy, "The House Public Works Committee," (Ph.D. Dissertation, University of Rochester, 1969), Murphy discusses the need of the committee to reach compromises in order to produce legislation and describes the operating style as "mutual advantage." Basically, committee members benefit when legislation is produced in the constituency-oriented committee and suffer when it is delayed.

56. David Zwick and Marcy Benstock, *Ralph Nader's Study Group Report on Water Pollution—Water Wasteland* (New York: Grossman Publishing, 1971).

57. Murray Edelman, *The Symbolic Uses of Politics* (Urbana: University of Illinois Press, 1967). Edelman discusses the quiescence of public opinion once a legislative victory is achieved, even when the victory brings little material change. We shall pursue this more fully in analyzing it as a strategy of the oil industry in both the issue areas.

58. Of course, power may be exercised in maintaining a non-decision status. And the industry did do things to insure the security of depletion. Since it was unopposed, it is hard to say whether this reflects industry power or the costs of organizing against the industry. The latter reflects a "mobilization of bias" rather than non-decision power.

59. Visibility of the issue differs from visibility of the decision making setting. In the cases under study the former refers to the fact that pollution is an easily seen problem compared to depletion allowance.

Notes—Chapter 5

1. Part of the reason for the low probability of success was that the amendments did not have the support of the Finance Committee or any of its senior members. Given further that changes in depletion were not in the House bill, such amendments, if they were accepted on the Senate floor, would not survive in conference. While the Finance Committee is not particularly successful on the Senate floor, the lack of success is usually on floor amendments favorable to special interests, not those opposed by such groups. At times these vehement anti-depletion Senators could not even get support for a roll call on depletion amendments.

2. According to a former Treasury official, percentage depletion in excess of cost claimed by corporate oil in 1968 was approximately $1.3 billion.

3. Schattschneider claims that "The biggest corporations in the country tend to avoid the arena in which pressure groups and lobbyists fight it out before congressional committees." And they resort to appeals to government only as a redress to a loss in intrabusiness conflict. (See E.E. Schattschneider, *The Semisovereign People* (New York: Holt, Rinehart and Winston, 1960), p. 41.) But winning early in the tax legislative process precludes this level of conflict; prior to 1969, the dispute had little more visibility than an intrabusiness quarrel.

4. Fred L. Zimmerman, "Cut in Oil Depletion Allowance May be Included in Tax Overhaul Measure House Unit is Preparing," *Wall Street Journal*, April 15, 1969, p. 3.

5. *Tax Reform Act of 1969*, Hearings before the Committee on Finance, U.S. Senate, 91st Congress, 1st Session on H.R. 13270, p. 4402.

6. *Congressional Record*, December 1, 1969, p. 36219.

7. Watts' and Ullman's opposition to the Morton motion was prompted, in part, because the motion called for proportional reductions in depletion for other mineral groups. They finally cast the deciding votes against the Morton motion, while Mills voted for it.

8. One former Treasury official, who later was a staff member to a Finance Committee Democrat, makes this claim. He argues that it was believed in Ways and Means that the industry was willing to accept 20 percent. He thought that Boggs was negotiating for "big oil." No one else, I interviewed, shared this interpretation of Boggs' action. I discount it for this reason and because it is inconsistent with other actions. If Boggs was on the side of "big oil," he could have voted for the Morton motion and settled for a reduction to 22 percent. (True, Mills then could have voted the other way.)

9. This further supports the idea that Boggs was trying to help smaller producers. When he originally offered his motion to cut depletion to 20 percent, it was combined with an increase of the percentage net limitation to 70 percent. Since majors usually reach 20 percent depletion before being subject to the 50 percent net limitation, an increase in the net limitation combined with a lowering of percentage depletion to 20 percent would assist the smaller producers.

10. Dole's statement is an exception to the rule for an oil state Senator. He is the only one who admitted publicly that a cut in depletion was acceptable to the industry.

11. *Congressional Record*, December 1, 1969, p. 36209.

12. The other items included: tax exempt interest, excluded 50 percent of net on long-term capital gain, appreciation on property contributed to charity to the extent deducted from income, and accelerated depreciation on real estate. H.R. 13270, 91st Congress, 1st Session, Paragraph 301, 1969.

The estimated value of the change appears in a memo from the Secretary of the Treasury to President Nixon on the subject of the Tax Reform Act of 1969, August 27, 1969, p. 4.

The inclusion of intangibles in the House bill LTP provision was an accident. George Bush was trying to eliminate it during an executive session of the Committee, when Hale Boggs wandered into the meeting from the House floor. Boggs asked what the discussion was about. He was told that it was on intangibles. He said he thought that they had decided to support the Treasury position. Bush asked a staff member how much money was involved. When told, he decided to give up the fight. Again, Boggs' action can only be explained in personal terms.

13. Mortimer Caplin, "Minimum Tax for Tax Preferences and Related Reforms Affecting High Income Individuals," *Indiana Legal Forum*, Vol. 4, Fall 1970, Number 1, pp. 89-91. Caplin covers in detail the development of the minimum tax provision.

14. *Congressional Record*, December 1, 1969, p. 36209. Later Dole, who

elicited a less than warm reaction from Senator Ellender for these remarks, reiterated his point but gave stronger support to Ellender's amendment to help gain "leverage" with the House in conference, pp. 36214-36215.

15. As we claimed earlier, the main benefit of this change would be for smaller producers.

16. Proxmire made the direct drilling proposal during his testimony before the Finance Committee. His proposal called for a direct subsidy of 25 percent of the intangible drilling costs on an exploratory well. This subsidy would be given in the form of a tax credit in addition to the expensing of intangibles on dry holes. There would be no limit on the amount of the subsidy in any one year, with an unlimited carry-forward period. To avoid the problem of the direct drilling subsidy's becoming a loophole, the proposal called for the credit also to be added to the taxable income of the taxpayer. The annual revenue loss to Treasury was estimated at $510 million as opposed to $795 million. The impact of the provision would be to assist smaller producers and wildcatters. Since it would eliminate intangibles for development wells, the major companies would suffer from the change. *Tax Reform Act of 1969*, Hearings before the Committee on Finance, U.S. Senate, 91st Congress, 1st Session on H.R. 13270, pp. 4210 and 4214.

17. This provision was dropped in conference.

18. Caplin, op. cit., p. 99. Percentage depletion on most other minerals was reduced by 1 percent.

19. Other measures of impact might be changes in production levels and new drilling operations. These are subject to many other factors aside from changes in the tax provisions and will not be investigated here.

20. Treasury memo, op. cit. This data was later made public in the Finance Committee hearings.

21. Small operators do not have many properties from which they can obtain greater than 20 percent depletion without first reaching the 50 percent net limitation. Nearly all of this change should have come from large producers.

22. This assumes a constant tax rate. The figure was derived from the following equation:

$$(1970 \text{ income} \times 1969 \text{ tax rate}) - 1969 \text{ actual taxes} =$$

tax increase from 1969 to 1970
with control for change in income

23. Thomas F. Field, "The Tax Treatment of Oil" (statement prepared for presentation before the Subcommittee on Priorities and Economy in Government), p. 11. Field represents Taxation with Representation and is responsible for development of this analysis. The impact of the price increase is on the small refiner who must pay more for crude and who does not have his own producing operations to which profits are transferred.

24. Whether Texaco did this in anticipation of a change in the depletion allowance is merely speculation, but the impact on depletion described is not. See *The New York Times*, Feb. 25, 1969, p. 53:3; Feb. 26, p. 58:3; and Feb. 27, p. 55:5 for information on acceptance of the increase.

25. Within two weeks Atlantic-Richfield, Skelly, Mobil, Sun, Humble, and Ashland announced similar price increases. See *The New York Times*, Nov. 12, p. 63:1; Nov. 16, p. 65:1; and Nov. 18, p. 71:8; Nov. 19, p. 67:3; Nov. 24, p. 55:3; and Nov. 25, p. 55:3.

26. William D. Smith, "Gulf to Increase Crude Oil Prices," *The New York Times*, November 12, 1970, p. 63:1. © 1970 by *The New York Times* Company. Reprinted by permission.

27. The Internal Revenue Service will not supply tax information on an individual or a corporation to check further on the impact of the depletion change. Annual reports issued by the companies are not helpful either. The financial data given stockholders about depletion differs from that in company tax returns. The former is closely related to what we would consider depreciation and is therefore meaningless for this analysis.

28. Most cited the 50 percent net limitation which prevents realization of 27½ percent on most operations.

29. It is uncertain whether Edelman explicitly states that reinforcement can occur. But from his analysis it seems that reinforcement is a logical possibility. If symbolic rewards are perceived as material by the recipients, then it is possible, and perhaps likely, that reinforcement of political action will result.

30. Both the McGovern and Wallace primary campaigns raised the tax reform issue, albeit from different perspectives. This is at least evidence that the candidates see it as a salient issue.

31. Frank V. Fowlkes and Harry Lenhart, Jr., "Two Money Committees Wield Power Differently," *The National Journal*, April 10, 1971. Bentsen was finally placed on Finance at the start of the 93rd Congress.

32. The water pollution control bills introduced in 1945 are as follows:

–H.R. 519 by Murdock (Dem., Ariz.) and companion S. 535 by Myers (Dem., Pa.).

–H.R. 587 by Smith (Rep., Mo.) and companion S. 330 by White and Brewster (Rep., Me.).

–H.R. 592 by Spence (Dem., Ky.) and companion S. 1037 by Barkley (Dem., Ky.).

–H.R. 4070 by Spence (Dem., Ky.) and companion S. 1462 by Barkley (Dem., Ky.).

33. Even the conservative Taft could be interested in expanding the federal government's authority when money for Ohio was involved.

34. *Congressional Quarterly Almanac*, 1948, p. 152.

35. M. Kent Jennings, "Legislative Politics and Water Pollution Control, 1956-1961" in Frederic N. Cleaveland (ed.), *Congress and Urban Problems* (Washington, D.C.: The Brookings Institution, 1969), p. 74.

36. A suit could only be filed by the Justice Department in cases of intrastate pollution when consent was granted by the state involved.

37. According to James Sundquist, even industry groups split on the bill. The Manufacturing Chemists' Association favored the Federal grant program of the bill. James L. Sundquist, *Politics and Policy* (Washington, D.C.: The Brookings Institution, 1968), p. 347. For the testimony by the Manufacturing Chemists' see *Federal Water Pollution Control*, Hearings before the House Public Works Committee, 87th Congress, I Session, March 15, 1961, p. 164.

38. The enforcement procedures had three steps: (1) conference, (2) hearing, and (3) then court action. They were pursued in that order, with the higher step only resorted to when a settlement was not obtained at the previous level.

39. *Water Quality Act of 1965*, Hearings before the Subcommittee on Air and Water Pollution of the Committee on Public Works, U.S. Senate, 89th Congress, 1st Session, on S. 4 (Jan. 18, 1965), pp. 29-37.

40. Ibid.

41. A first attempt to give HEW standard setting powers was made during the 88th Congress. But by the time the House committee reported the bill, it was too late for floor action. See Sundquist, op. cit., pp. 350-351, for a description of activity.

42. *Water Pollution Control Hearings on Water Quality Act of 1965*, Hearings before the House Public Works Committee, 89th Congress, 1st Session (1965), p. 192.

43. Ibid., p. 194.

44. Sundquist, op. cit., p. 364.

45. Ibid.

46. William M. Blair, "Water Pollution Fight Called Too Slow," *The New York Times*, May 26, 1971, p. 68:6.

47. While the implications of the Wright amendment may seem obvious, the first new reports of the conference agreement did not find this flaw. See *The New York Times*, October 14, 1966, p. 1:6.

48. *Water Pollution Control–1966*, Hearings before the Subcommittee on Air and Water Pollution of the Senate Public Works Committee, 89th Congress, 2nd Session (April 1966), p. 258.

49. Sundquist, op. cit., p. 365.

50. *The New York Times*, May 27, 1970, p. 26:3. © 1970 by The New York Times Company. Reprinted by permission.

51. Ibid., April 11, 1967, p. 20:1; April 12, p. 33:1; April 13, p. 52:7; and July 28, p. 62.

52. Ibid., April 16, 1967, p. 41:1.

53. Between July 12, when *The New York Times* reported a vessel spill off

Liverpool, England, and November 14, when there was a Hess Oil Company barge spill, no major spills occurred.

54. *The New York Times*, February 15, 1970, p. 2:3. With tankers of over 10 million gallon capacity, even the $14 million liability in the Senate bill would easily be exceeded.

55. E.W. Kenworthy, "U.S. Outlines Plan for Oil Cleanup," *The New York Times*, June 2, 1970, p. 28:4.

56. *The New York Times*, July 25, 1970, p. 13:1.

57. William M. Blair, "Muskie Attacks Oil Spill Rules," *The New York Times*, August 5, 1970, p. 42:1.

58. E.W. Kenworthy, "Revision Delayed on Oil Spill Code," *The New York Times*, December 1, 1970, p. 1:7.

59. *Code of Federal Regulations* (Washington, D.C.: G.S.A., January 1, 1972), Vol. 40, p. 291.

60. *Water Pollution Control Programs*, Hearings before the Subcommittee on Air and Water Pollution of the Committee on Public Works, U.S. Senate, 92nd Congress, 1st Session (February 1971), p. 764.

61. Ibid., p. 762.

62. All reporting prior to November 21 was voluntary. Ibid., p. 764.

63. *The New York Times*, March 24, 1970, p. 27:1.

64. In all, twelve wells were out of control at one time or another over a three-month period.

65. *The New York Times*, May 27, 1970, p. 26:3.

66. Ibid., December 3, 1970, p. 25:1.

67. The exact provision from Section 407 of the Rivers and Harbors Act is as follows:

§ 407. Deposit of refuse in navigable waters generally.

It shall not be lawful to throw, discharge, or deposit, or cause, suffer, or procure to be thrown, discharged, or deposited either from or out of any ship, barge, or other floating craft of any kind, or from the shore, wharf, manufacturing establishment, or mill of any kind, any refuse matter of any kind or description whatever other than that flowing from streets and sewers and passing therefrom in a liquid state, into any navigable water of the United States, or into any tributary of any navigable water from which the same shall float or be washed into such navigable water; and it shall not be lawful to deposit, or cause, suffer, or procure to be deposited material of any kind in any place on the bank of any navigable water, or on the bank of any tributary of any navigable water, where the same shall be liable to be washed into such navigable water, either by ordinary or high tides, or by storms or floods, or otherwise, whereby navigation shall or may be impeded or obstructed: *Provided*, That nothing herein contained shall extend to, apply to, or prohibit the operations in connection with the improvement of navigable waters or construction of public works, considered necessary and proper by the United States officers supervising such improvement or public work: *And provided further*, That the Secretary of the Army, whenever in the judgment of the Chief of Engineers anchorage and navigation will not be injured thereby, may permit the deposit of any material above

mentioned in navigable waters, within limits to be defined and under conditions to be prescribed by him, provided application is made to him prior to depositing such material; and whenever any permit is so granted the conditions thereof shall be strictly complied with, and any violation thereof shall be unlawful.
Mar. 3, 1899, c. 425, § 13,30 Stat. 1152.

68. Ibid., Paragraph 411.

69. *Water Pollution Control Programs*, op. cit., p. 78. The totals of the columns are not correct, but are given as reported in the Senate hearings.

70. Actually, all three stages involve material consequences. Since both the non-decision and symbolic phases maintain the status quo, they maintain the industry position and therefore have consequences. The stages are differentiated rather on the basis of perceived and actual material changes. In both the non-decision and material stage the perception of both the industry and the opposition are the same and are in fact congruent with the change (or lack of change). Only in the symbolic stage do the groups perceive differently. There is action in the form of legislation but no real impact. The organized industry group is aware of this, but the opposition may be fooled.

71. Sundquist, op. cit.

72. Waggonner has been described by a Committee member as the "only real reactionary" on the Committee.

73. They claimed that Long worked to get Nelson on the Committee in order to block Walter Mondale of Minnesota. While this may be true, it is difficult to believe that Long would not have preferred Bentsen to Nelson. Given the close vote of the Steering Committee, it becomes even more difficult to accept.

74. Cohen indicated during the Ways and Means hearings on the impact of the 1969 tax reforms that the Administration opposed any further major tax revision. *The New York Times*, May 2, 1972, p. 47:6. Two days previous to that, speaking before the Federal Tax Institute of New England, Cohen claimed that the oil industry was paying $600 million a year in additional federal taxes, and that this was mainly due to the change in the depletion allowance. (The evidence in this study clearly disputes Cohen's statement. The industry's larger tax payments were in fact a result of higher profits.) He then concluded that "with the energy shortage that is facing us and the dire need for a coordinated energy supply, we should be sure we move cautiously and intelligently with a coordinated energy resource program." *The New York Times*, April 30, 1972, p. 25:3. © 1972 by The New York Times Company. Reprinted by permission.

75. "Big Oil Changing Image to Anti-pollution 'Zealots,' " *The Boston Globe*, October 24, 1971, p. 2:2.

76. *The Washington Post*, November 26, 1971, p. 24:1.

77. Gladwin Hill, "Santa Barbara, 2 Years After Its Oil Well Blowout, Still Hopes to Curb Offshore Drilling," *The New York Times*, January 27, 1971, p. 16.

78. *The Washington Post*, December 5, 1971, p. E14:3 and *The Boston Globe*, December 19, 1971, p. 57:1.

79. Elsie Carper, "Kennedy, Morton Clash on Offshore Oil," *The Washington Post*, December 9, 1971, p. A2. Needless to say, we have not mentioned the clashes between the industry and environmental groups over the proposed Alaskan pipeline. This is another study in and of itself.

80. Richard T. Cooper, "Oil Firms' World Profits Soared in '73, Senate Finds," *The Boston Globe*, January 23, 1974, p. 1:1.

81. Schattschneider, op. cit., p. 74.

82. The Executive agencies offer new sources of low profile decision-making at which we might expect the industry to be more successful.

83. Raymond Bauer, et al., *American Business and Public Policy* (New York: Atherton Press, 1963), immediately comes to mind.

Notes–Appendix

1. Most of the legislative action on the Water Quality Improvement Act occurred in 1969. The House-Senate conference committee on the bill did not reach final agreement until early 1970 and President Nixon signed the bill in March 1970. Throughout this study when I mention the 1969 bill, I mean the Water Quality Improvement Act. I believe that the 1969 date is more accurate when discussing the legislation.

2. I avoided scheduling more than two interviews on any one day. Since I rarely wanted to cut off an interview, I did not schedule them back to back. In addition, I felt it desirable to write up one interview before starting another. A further problem was created in lag times between first contacting an interviewee and the actual interview. One oil lobbyist vital to the study had just left for Europe when I first contacted his office. I finally spoke with him at length two months later.

3. In this regard I disagree with the advice offered by Lewis A. Dexter, *Elite and Specialized Interviewing* (Evanston: Northwestern University Press, 1970), pp. 59-60. He believes that the use of a tape recorder is advisable, but considers only informational gains and not the losses I mention. Dexter holds what I believe is a minority opinion among students of Congress on this point. However, I agree with many of the other suggestions he offers, especially when he says that elite interviews should be a form of "discussion rather than rat-a-tat-tat questioning."

Index

Index

198

About the Author

Bruce Ian Oppenheimer is Assistant Professor of Politics at Brandeis University. He received the B.A. from Tufts University and the M.A. and the Ph.D. from the University of Wisconsin. In 1970/71, Professor Oppenheimer was a graduate fellow in Governmental Studies at The Brookings Institution, where he began research on interest groups in the Congressional process. He has been selected as an American Political Science Association Congressional Fellow for 1974/75.